CASTABLE POLYURETHANE ELASTOMERS

SECOND EDITION

CASTABLE POLYURETHANE ELASTOMERS

SECOND EDITION

I.R. CLEMITSON

CRC Press
Taylor & Francis Group
Boca Raton London New York

CRC Press is an imprint of the
Taylor & Francis Group, an **informa** business

CRC Press
Taylor & Francis Group
6000 Broken Sound Parkway NW, Suite 300
Boca Raton, FL 33487-2742

First issued in paperback 2017

ISBN-13: 978-1-4987-2637-5 (hbk)
ISBN-13: 978-1-138-80920-8 (pbk)

Library of Congress Cataloging-in-Publication Data

Clemitson, Ian.
  Castable polyurethane elastomers / author, Ian Rex Clemitson. -- Second edition.
      pages cm
  Includes bibliographical references and index.
  ISBN 978-1-4987-2637-5 (hardcover : alk. paper) 1. Polyurethane elastomers. 2. Polyurethanes--Industrial applications. 3. Plastics--Molding. I. Title.

TP1180.P8C54 2015
678'.72--dc23   2015014854

Visit the Taylor & Francis Web site at
http://www.taylorandfrancis.com

and the CRC Press Web site at
http://www.crcpress.com

Contents

IV Properties

Preface

Castable Polyurethane Elastomers is a practical guide to the production of castable polyurethane articles. These articles can be as simple as a doorstop to items used in nuclear and military industries. The book shows the progression from the raw materials needed to produce prepolymers to the production of prepolymers. This includes both the chemistry and the practical side of the production processes.

The production of polyurethane components is explained from both the theoretical and practical aspects covering the different types of systems available and the reasons for choosing the right system, on both the micro and macro levels. Curing and post-curing operations are also covered.

The use of the traditionally quoted properties, for example, tensile strength and hardness, are often not the best for selecting the correct system to use. In the section on properties the importance of using the correct property for the application is explained. The application of polyurethanes in various fields is expanded on and the logic for suitability discussed. The effect of changes to the original application details is dealt with.

As the world is not a perfect place there are sections on problem solving and possible solutions. Throughout the book there is an emphasis on the health and safety aspects that should be observed at all times.

The book is aimed at people entering the polyurethane world and those who work with it.

This second edition details newer methods and standards in the processing of castable polyurethanes. The importance of health and safety is emphasized with the chapter on this brought forward in the book.

The Author

Ian Clemitson has worked in the chemical industry for forty-five years. He worked predominantly in the polymer industry, both in production and development, concentrating on rubbers and polyurethanes. In 1999 he earned a masters degree from the University of Technology in Sydney, Australia. His thesis was titled "The Influence of Polyurethane Chemistry on Erosive Wear." During this time he was working at the R & D laboratory of Warman International in Sydney doing research into polyurethanes and other elastomers.

In the concluding years of his fulltime work he was employed in the development and manufacture of polyurethane elastomers and foams. He also gave lectures at the local training institute on aspects of polyurethane and rubber technology.

He has authored two books on polyurethane elastomers, namely, *Castable Polyurethane Elastomers* and *Polyurethane Casting Primer*, both published by CRC Press.

Symbols and Acronyms

Symbol	Description
ASTM	American Standard Testing Methods
°C	Celsius (Centigrade)
E @ B	Elongation at break
E100	Ethacure 100
E300	Ethacure 300
FDA	Food and Drug Administration
ISO	International Standard Organization
MAK	Inhalable fraction
MbOCA	See MOCA
MDI	Methylene diisocyanate
MOCA	Methylene bis-orthochloro-aniline
Mod	Modulus
NDI	Naphthalene diisocyanate
PEL	Permissible exposure limit
PTMEG	Polytetramethylene glycol
Rx	A chemical group
SDS	Safety Data Sheet
TDI	Toluene diisocyanate
TLV	Threshold limit value
TS	Tensile strength
TWA	Time weighted average

Part I

Introduction to Polyurethanes

1

Introduction to Polyurethanes

1.1 Background

Castable polyurethanes are a group of organic polymers that derive their name from the presence of the urethane bond in their structure. This group is formed by the reaction of an isocyanate with a hydroxyl group of a macro diol. In order to develop longer chains, diols are used to further extend the chains. To obtain specific properties isocyanates and hydroxyl groups of higher functionalities can also be used.

$$-R_1-N=C=O \quad + \quad H_2O-R_2- \quad \longrightarrow \quad -R_1-NH-\underset{O}{\overset{\overset{\displaystyle O}{\|}}{C}}-O-R_2-$$

$$\text{Isocyanate} \qquad\qquad \text{Hydroxyl} \qquad\qquad\qquad\qquad \text{Urethane}$$

Diamines are also used to chain extend the initial polyurethane product. The reaction of the amine and the isocyanate group forms a urea bond. A poly(urea urethane) is formed. Commonly it is still referred to as polyurethane.

$$-R_1-N=C=O \quad + \quad H_2R-R_2- \quad \longrightarrow \quad -R_1-NH-\underset{O}{\overset{\overset{\displaystyle O}{\|}}{C}}-NH-R_2-$$

$$\text{Isocyanate} \qquad\qquad \text{Amine} \qquad\qquad\qquad\qquad \text{Urea-urethane}$$

A closely related series of compounds are the polyureas. These are made from the reaction of diisocyanates with a polyamine. The reactions are very much more vigorous than those using polyols.

1.2 History

Isocyanates were first isolated by Wurtz in 1849 when he prepared isocyanates by the reaction of cyanates with organic sulfates. In 1850 Hoffman first prepared aromatic isocyanates.

In 1937 Du Pont introduced nylon, a polymer closely related to polyurethanes. During the 1930s Otto Bayer and his team developed the first commercial polyurethanes based around the reaction of polyols and diisocyanates. (German Patent 728.981 (1937)). He worked at I.G. Farben in Leverkusen where he was exposed to completely new fields of research, such as rubber chemistry, pharmaceutical research and crop protection. Otto Bayer's greatest achievement was ultimately the invention of polyurethane chemistry. The principle of polyaddition[1] using diisocyanates is based on his research.

Although the production of macromolecular structures was already a line of research that held promise for the future, Otto Bayer's basic idea of mixing small volumes of chemical substances together to obtain dry foam materials was seen as unrealistic. But after numerous technical difficulties, Bayer eventually succeeded in synthesizing polyurethane foam. It was to take ten more years of development work before customized materials could be manufactured on the basis of his invention.

$$-\overset{O}{\overset{\|}{C}}-(CH_2)_4-\overset{O}{\overset{\|}{C}}-\underset{H}{\overset{}{N}}-(CH_2)_6-\underset{H}{\overset{}{N}}-\overset{O}{\overset{\|}{C}}-(CH_2)_4-\overset{O}{\overset{\|}{C}}-\underset{H}{\overset{}{N}}-(CH_2)_6-\underset{H}{\overset{}{N}}-$$

Basic Nylon 66 formula

During the Second World War, Otto Bayer's polyurethanes were developed into foams and coatings for use in the German military. The next major step was the patents issued to William Hanford and Donald Holmes of E.I. du Pont de Nemours & Company, for a process of making and processing castable polyurethanes. The patent titled "Process for Making Polymeric Products and for Modifying Polymeric Products Polyurethane," number 2,284,896, was granted in 1942.

The first commercial polyurethanes were introduced in 1952 and were based on polytetramethylene glycol (PTMEG) and aromatic diisocyanates. The industry experienced a strong surge forward when cheaper polyols based on polypropylene glycol came onto the market. This gave a big boost to the foam market.

1.2.1 Timeline

The commercial development of polyurethane gained momentum from just before the Second World War. The development split in two during the war with separate strands in Germany and America.

1849 Wurtz, Inorganic isocyanates.

1869 Gatier, Aliphatic isocyanates.

---

[1] The current explanation is that the reaction is a "step-wise" condendation polymerization.

1884 Hentschel developed method to produce isocyanate by action of phosgene on amines.

1937 Carothers (DuPont) patented super polyamides (Nylon). Nylon 66 was developed in 1935 and patented and commercialized in 1937.

1937 Professor Otto Bayer's reaction of diisocyanate with amines.

1937 Professor Otto Bayer's reaction of 1,4-butylene glycol with diisocyanate was used to produce bristles.

1938 Reike et al., Linear polyurethanes from glycols and diisocyanate US Patent (issued in 1950) 2,511,544 (Alien Property custodian).

1940 Schlack reacted polyesters with diisocyanates such as 1,6-hexamethylene diisocyanate. Products had low softening temperatures. German Patent Appl. J-66,330.

1940 Christ and Hanford, Formation of elastic urethane products, US Patent 2,333,639.

1941 Diisocyanates and dihydroxy compounds, Lieser, US Patent 2,266,277.

1942 Hexamethylene diisocyanates with alkyd resins, US Patent 2,282,827.

1942 Catlin was granted a broad patent, US Patent 2,284,637, covering the reaction of glycols and diisocyanate fiber and flexible sheets.

1942 Hanford and Holmes, Reaction of polyfunctional compounds containing active hydrogens, US Patent 2,284,896.

1945 Hoff and Wicker, Alternate route for production using glycol bis(chloroformates) and amines (Perlon U).

1950 Commercial production of toluene diisocyanate.

1950 Bayer introduces Vulkalon rubbers. Unstable prepolymer system based on polyester and NDI.

1950 ICI England introduces Vulkaprene A based on polyesteramides.

1952 Cheaper polyethers based on propylene oxide introduced in the United States.

1952 Cheaper polyethers based on castor oil introduced in the United States.

1953 Goodyear Tire and Rubber Company introduced Chemigum based on using a deficit of isocyanates.

1956 DuPont introduced Adiprene based on PTMEG.

1957 Polyurethane-grade polyethers based on propylene oxide introduced.

1964 Carbonate-based polyesters patented (US Patent 3133113).

1970 Clear, light-stable polyurethanes introduced.

1981 Ethylene oxide (EO) tipped PPG came on the market.

1987 PPG produced using double metal catalyst (DMC) – Polyols more linear and higher diol content.

1.2.2   Nomenclature

Various groups in the chemistry field have different names for the same chemical group. The following table illustrates some of the groups.

	Urethane	Urea	Amide
Formula	-NH-CO-O-	-NH-CO-N-	-CO-NH-
Organic Chemistry	Carbamate	Urea	Amide
Polymer	Urethane	Urea	Nylon
Biological	Not applicable	Urea	Peptide

1.2.3   Nature of Polymer

Depending on the ratios of the reacting agents, the final product can vary from a thermoplastic polymer through being a pseudo thermoset to a thermoset polymer. If the ratio of polyol to diisocyanate is designed to give a material that is hydroxyl ended, a thermoplastic material is obtained. These polymers can be processed in the same type of extrusion and injection molding machines as other thermoplastics such as polyethylene, polypropylene, PVC, etc. The formation of hydrogen bonding between the polymer chains gives the final product its mechanical strength.

The reactant ratios can be adjusted so that the ends of the chains can be further chain extended with either a diol or amine curative. This can be achieved by having an initial isocyanate-to-macro diol ratio of 2 to 1. This is the most popular route for handling castable polyurethanes.

An early modification was to introduce some double bonds into the polymer so that cross-linking can be obtained either with free radicals or with sulfur. The advantage of this system is that the polyurethane can be processed and molded using conventional rubber processing machinery.

1.2.4   Applications

Polyurethanes find many applications in the industrial, domestic, medical and military fields. The applications are not always as obvious as other polymers (e.g., PET bottles) but can be found in uses such as sliding gate wheels. Using the CASE[2] system of classification of polymers, polyurethanes are found in all four categories. Polyurethanes can be tailored to suit a wide range of needs by changes to the chemistry.

Coatings

Polyurethanes can be used in both industrial and domestic coatings. There are a number of different chemistries that can be employed to give excellent results. The simplest is the use of atmospheric moisture to cure the polyurethane. One of the earliest uses were coatings prepared from castor oil prepolymers that were coated and then allowed to cure from the moisture in the air and the under surface. The initial viscosity was low to allow for the escape of the carbon dioxide released and for the basic leveling of the coating.

For harder and tougher polyurethane coatings that offer both chemical and abrasion resistance, single and two pack systems are available. The two pack system will provide a hard glossy finish with all the desired properties. Due to the fact that the spray has a high level of unreacted isocyanate in an aerosol form, great care must be taken with industrial hygiene. These systems can be further modified by using a blocked isocyanate so that the sprayed-on coating can be cured by the application of heat.

Coating of pipes and launders in the mining industry is an important use for polyurethane coatings.

Aliphatic isocyanates can be used where yellowing is a disadvantage in the finished product.

By producing systems that contain only urea bonds instead of urethane bonds (by using polyamines instead of polyol), very fast curing coatings can be prepared with excellent properties without the use of catalysts or heat to fully cure them.

Adhesives

Polyurethanes can be designed to give a very fast setup with good green strength and grab properties. This makes them ideal in situations when the various components can be joined together without the need for jigs and clamps, thus making the processing costs lower. Modified isocyanates can be used to provide good interlayer adhesion. With the use of proprietary polyurethane-based bonding agents, polyurethane can be bonded to glass.

Polyurethane systems with a low degree of blowing can be used in the adhesion of metal plates to an underlay such as wood. The foaming will fill in any unevenness in the surface. A similar application is attaching a new liner to an old enamel bath.

The toughness, ability to flex and strength of polyurethanes are important in the use of polyurethane as an adhesive.

Hydrolysis and chemical resistance are important and can be used

---

[2]**C**oatings, **A**dhesives, **S**ealants and **E**lastomers.

to advantage in certain situations where long-term integrity is required. Polyurethane conveyor belts can be spliced using a polyurethane adhesive with the same basic chemistry. The ends of the belts are cleaned and roughened and a bonding agent is applied prior to the use of the adhesive.

Sealants

Polyurethanes are used as sealants in both the solid form and as gap filling foams. A major use of polyurethane sealants is to provide moisture barriers in highrise buildings. They are used between the concrete and the floor surface to provide a waterproof barrier. They are also used in areas such as elevator(lift) wells, patios and planter boxes to control the movement of moisture.

Polyurethane sealants used in waterproofing are designed to have the following major properties:

- Sufficient hydrolytic stability to last the life of the structure
- Sufficient stretch and flexibility to withstand normal movement in the concrete
- Ability to bridge small gaps
- Ability to form a continuous layer with no holes
- Ability to bind well to concrete
- Sufficient thixotropy to allow vertical coating

In situ expanding polyurethane foam can be used to seal gaps in construction work. Solid polyurethane can be used to provide seals in pipe work.

1.2.5   "CASE" Elastomers

Elastomers form the largest group in the polyurethane family and can be further divided into a number of distinct groups.

Fibers

The initial driving force resulting in the development of polyurethanes was the commercial need to find an alternative to Nylon prior to World War II. Research by the Bayer group resulted in the team led by Otto Bayer obtaining patents for polyurethane fibers and foams. The most successful polyurethane fibers are Perlon and spandex.

Millable Polyurethanes

Early forms of polyurethane elastomers were developed to enable processing by rubber machinery. This included the incorporation of fillers and curatives in both open mill and in intensive mixers such as the Banbury mixers. The molding is carried out in conventional rubber molds and presses. Polyester

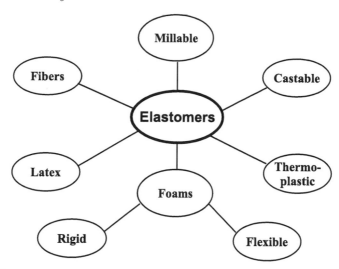

FIGURE 1.1
Polyurethane elastomer relationships.

and to a lesser extent polyether types are available in varieties that can either be peroxide or sulfur cured.

The advantage of this system is that normal rubber processors can produce polyurethane parts with wear and chemical resistance using their existing capital equipment.

Castable Polyurethane Elastomers

Castable polyurethanes are one of the three major segments of the polyurethane industry. The nature of the process allows business to start up with a very low capital outlay, progressing to fully automated systems when required.

The process consists of a few basic steps:

- Dispense two to five "dry" ingredients at the required temperature.
- Mix completely.
- Cast into preheated — prepared mold.
- Cure fully.
- Demold.
- Trim excess flash.

Polyurethane itself can be used to make the molds for short to medium production runs. The degree of complexity of the whole process is a function of the end use of the product. The specified use of the product will require the selection of the type and chemistry of the polyurethane and the quality of the mold used. A typical example is if there is food contact involved; most amine curatives cannot be used so MDI/ hydroxyl systems are more commonly employed.

The uses of castable polyurethanes in the domestic field are not directly noticeable. However there are several everyday examples. Suspension bushes used in off-road vehicles can be made from polyurethanes. Boat trailer rollers as well as applications in engine mounts and general finishing of crafts of all sizes, are very important uses of polyurethanes in the maritime industry.

In the garden the molds used for making pavers are often made from polyurethane. Molded sheets of polyurethane can be used to make a stamp to produce designs on concrete driveways. A similar method is used in the production of sound-absorbing panels on the side of the highway. The technology of polyurethane wheels has developed to a point where they can be supplied as a spare wheel for motor cars.

Industrial applications of polyurethanes are quite varied and often limited only by the imagination of the applications engineer. Polyurethanes find use as an engineering material in its own right but can also be used as a prototype material to allow production of parts on a low volume basis prior to large-scale production using expensive injection molding tools.

With the wide scope of variations in chemistry the most suitable polymer system for various applications can be selected. Different systems need to be selected for shock-absorbing applications as opposed to applications where the maximum immediate return of energy is required. Industrial rollers of all sizes are an important field. The scope ranges from printing rollers to rollers for large mining mills. Wheels can range from miniature parts to forklift wheels. The mining industry provides many uses for polyurethanes. Typical applications include

- Pumps:

    - Liners
    - Throats
    - Seals

- Cyclones:

    - Integral bodies
    - Liners
    - Spigots (under and overflow)

- Conveyor systems:

    - Belts
    - Head and drive drums
    - Idler wheels
    - Scrapers

- Beneficiation:

- Mill liners
- Lifter bars
- Screens
- Roller wheels

There are many other examples in general industry where either the chemical or wear resistance makes polyurethane a more economical choice compared to other polymers or to metals and ceramics. Polyurethanes also find use in the nuclear industry due to their resistance to radiation. Medical applications of polyurethanes can be found in diverse applications such as catheters and even parts made from scans to make three-dimensional models of tumors in the human body.

Military applications include items such as binders for rocket oxidants and for coatings on rockets.

Foams

Polyurethane foams rely on an area of urethane chemistry different from that used in solid elastomers where in many cases maximum strength properties are required. Foams constitute one of the major segments of polyurethane use and tend to drive the search for different polyols.

There are several distinct segments of foam technology and types of foam:

- Rigid foams:

  - Insulation
  - Void filling, e.g., surfboards

- Flexible foams:

  - Cushions and mattresses
  - Integral skinned, e.g., for car dashboards
  - Underlays for carpets

- Void filling:

  - Rock stabilization

The chemistry of the system controls the type of foam that is produced. The networks are more highly cross-linked by using multifunctional polyols. The foaming of the system requires a blowing agent, foam stabilizing surfactants and catalysts to control the speed of reaction. The blowing agents used can vary from water (to develop CO_2 internally) to low-boiling liquids. The choice of these liquids is controlled. Some materials that were very popular (e.g., CFCs) are potential greenhouse gas hazards.

Foams can be produced containing fire-resistant materials or they can be made to contain isocyanurate bonds that give a higher degree of fire resistance.

Latex

Polyurethanes can be processed as an emulsion to form completely extended polyurethane polymers. On coagulation and removal of the suspending medium, the polyurethane can be formed into the desired state normally as a thin layer.

If the latex is in an aqueous solution, coatings can be made that do not have any VOC problems.

Thermoplastic Polyurethanes (TPUs)

Thermoplastic polyurethanes differ from castable polyurethanes in that they can be melted without undue degradation. This means that they can be processed in standard-style plastic machinery such as injection molding, extruders and blow molding machines.

The polyurethanes are produced with different polyol-to-isocyanate ratios from the castable materials. The ratios used give materials terminating with hydroxyl groups as opposed to the NCO of the castable polyurethane group.

These ranges of materials are particularly suited for long production runs of precision parts. The parts normally have a reasonably thin cross-section, for example, shock absorber boots. The ability to be extruded allows the production of polyurethane tubing and the coating of wires with polyurethane.

2

Chemistry of Polyurethanes

2.1 Introduction

2.1.1 Background

In 1849 Wurtz discovered the formation of aliphatic isocyanates when he re-acted organic sulfates with cyanates. This represents the first recorded exper-imental work leading ultimately to the preparation of urethanes.

$$R_2SO_4 \quad + \quad 2KCNO \quad \longrightarrow \quad 2RNCO \quad + \quad K_2SO_4$$

The first important commercial development was due to the work of Pro-fessor Otto Bayer in 1937, who discovered how to make a polymer using di-isocyanates employing an "addition polymerization" technique when working on a polymer fiber to compete with nylon. Initially the development was con-sidered as impractical. In 1938 Rinke and associates succeeded in producing a low-viscosity melt that could be formed into fibers. This led to the production of many different types of polyurethanes.

Polymer chemistry developed a nomenclature (language) different from the classical organic chemistry. The group "-NHCO-" in polymer chemistry is known as nylon whereas in biochemistry it is known as a peptide and in pure organic chemistry as an amide group. The urethane group "-NHCOO-" is called a carbamate in organic chemistry.

Polyurethanes are named because of the presence of the urethane linkage, which is illustrated below:

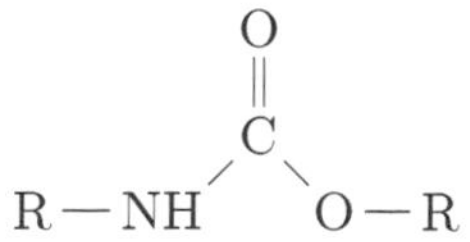

$$
\begin{array}{c}
O \\
\parallel \\
C \\
R-NH \diagup \quad \diagdown O-R
\end{array}
$$

Urethane linkage

Intensive research determined that there were many modifications that could be made to the chemistry surrounding the urethane linkage. During the 1940s and early 1950s Du Pont and ICI developed castable polyurethanes.

Wright and Cummins [12] classified polyurethanes into eight different clas-sifications:

1. Linear polyurethanes

2. Castable polyurethanes

3. Millable polyurethanes

4. Thermoplastic polyurethanes

5. Cellular polyurethanes

6. Sprayable polyurethanes

7. Porometric polyurethanes

8. Spandex fibers

Of these groups, the castable, thermoplastic and sprayable polyurethanes have the greatest abrasive and erosive wear resistance.

Within each family there are various factors that give it particular properties and make it most suitable for various uses. The major factors are the degree of branching or cross-linking. This linking of the chains may either be permanent or by hydrogen bonding. This cross-linking interacts with the chain stiffness and the crystallinity of the polyurethane. Cast polyurethanes have a moderate degree of cross-linking, in most cases due to interchain attractions. Rigid foams have high chain stiffness and a high level of branching and cross-linking. Millable polyurethanes, by contrast, have a limited degree of cross-linking and low chain stiffness, even lower than thermoplastic polyurethanes.

The castable and sprayable types are commercially the most popular. They are easier to process, and parts can be made with an initially low capital expenditure.

The millable materials have a diene group (-CH=CH-) included in the polymer to allow for cross-linking by sulfur or by peroxides. However, the capital costs to use millable polyurethanes are high because rubber mills, presses and heavy-duty molds are needed to produce an article.

A variety of polymeric sub-units are used to make polyurethanes. These include polyesters and polyethers. The major interchain linkages are molecular forces such as hydrogen bonding and the London force. Depending on the type of chain extender and processing temperature, there may also be biuret or allophanate cross-links.

Polyesters make tough and wear-resistant urethanes. The one major drawback is hydrolysis at the ester grouping. The hydrolysis can either be acid or alkali promoted. In more neutral conditions the major breakdown product is normally adipic acid that then catalyzes further attack. The normal approach is to use carbodiimides to block further breakdown. Polyols based on polypropylene carbonate produce polyurethanes with polyester characteristics. The modified ester structure produced gives enhanced hydrolysis resistance.

Polyethers are not as tough as polyester urethanes. They do, however, provide a far greater degree of hydrolysis resistance due to the presence of an ether group instead of an ester grouping in the backbone.

2.1.2 Basic Reactions of Urethanes

This book studies the castable family of polyurethanes. The cured polyurethane elastomer is made from three main ingredients:

1. Polyol
2. Isocyanate
3. Chain extender

The urethane linkage is the fundamental group in polyurethane chemistry. The main reaction in the creation of castable polyurethanes is that of a hydroxyl with an isocyanate:

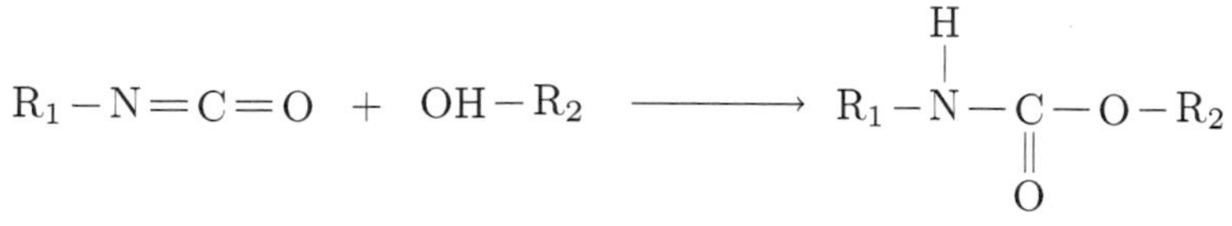

$$R_1-N=C=O \ + \ OH-R_2 \longrightarrow R_1-N-C-O-R_2$$

Isocyanate   Hydroxyl   Urethane unit

The next most important reaction is that of the amine group with an isocyanate:

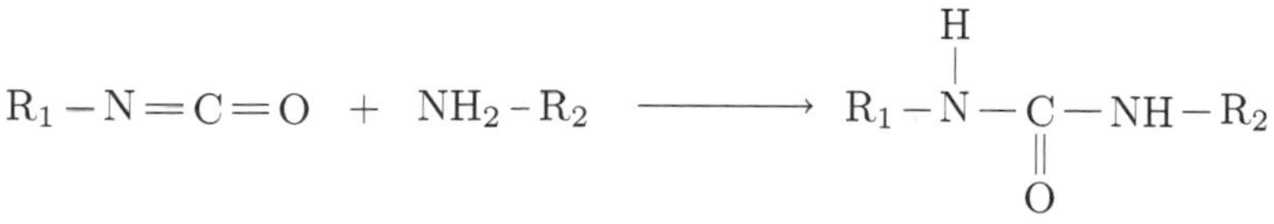

$$R_1-N=C=O \ + \ NH_2-R_2 \longrightarrow R_1-N-C-NH-R_2$$

Isocyanate   Amine   Urea unit

The new compound formed does not have groups available for the reaction to be repeated to form a polymer. The simplest method is to use a difunctional hydroxyl and isocyanate. This will allow for the reaction to continue and a polymer to form. The most basic is the formation of polyurethane from a diol and a diisocyanate:

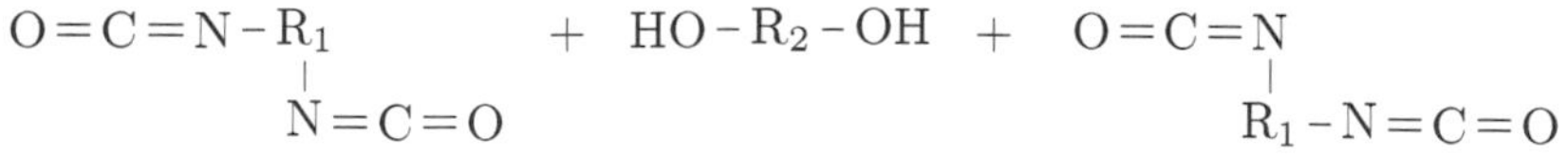

$$O=C=N-R_1 \ + \ HO-R_2-OH \ + \ O=C=N$$
$$N=C=O \qquad\qquad\qquad\qquad\qquad R_1-N=C=O$$

Diisocyanate   Diol   Diisocyanate

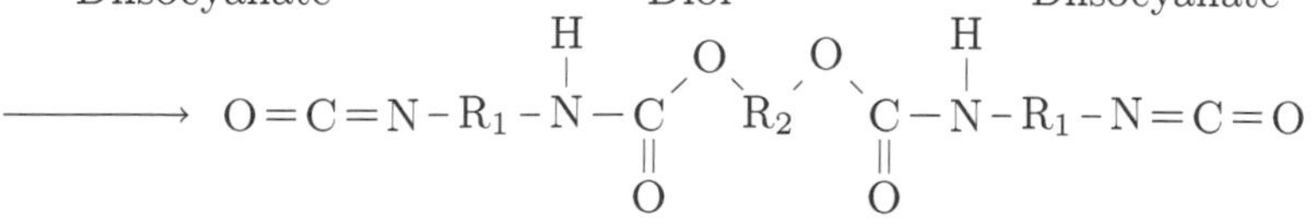

$$\longrightarrow O=C=N-R_1-N-C \underset{O}{\overset{O}{R_2}} C-N-R_1-N=C=O$$

Polyurethane prepolymer

There are advantages in speed of reaction by replacing the diol with a diamine. In this case the reaction between the diisocyanate and the diamine forms a urea group. This reaction is important in both chain extension and another polymer, a close relative of polyurethanes, namely polyureas.

$$N=C=O \qquad\qquad\qquad\qquad N=C=O$$
$$|\qquad\qquad\qquad\qquad\qquad\qquad |$$
$$O=C=N-R_1 \qquad + \quad NH_2-R_2-NH_2 \quad + \quad O=C=N-R_1$$

Diisocyanate   Diamine   Diisocyanate

$$\qquad\qquad\qquad\qquad\qquad\qquad H \qquad\qquad\qquad\qquad H$$
$$\qquad\qquad\qquad\qquad\qquad\qquad | \quad NH \quad NH \quad |$$
$$\longrightarrow \quad O=C=N-R_1-N-C \qquad R_2 \qquad C-N-R_1-N=C=O$$
$$\qquad\qquad\qquad\qquad\qquad\qquad || \qquad\qquad\qquad ||$$
$$\qquad\qquad\qquad\qquad\qquad\qquad O \qquad\qquad\qquad O$$

Polyurea prepolymer

The nature of the polymerization is best described by a step growth route. The reactants join together by the rearrangement of terminal atoms:

$$\qquad\qquad\qquad\qquad\qquad\qquad\qquad\qquad\qquad\qquad H$$
$$\qquad\qquad\qquad\qquad\qquad\qquad\qquad\qquad\qquad\qquad |$$
$$R_1-N=C=O \;+\; H-O-R_2 \quad\longrightarrow\quad R_1-N-C-O-R_2$$
$$\qquad\qquad\qquad\qquad\qquad\qquad\qquad\qquad\qquad\qquad ||$$
$$\qquad\qquad\qquad\qquad\qquad\qquad\qquad\qquad\qquad\qquad O$$

The nature of the reaction, though similar to "addition polymerization," is often classified as condensation polymerization as two different monomers are involved. (There are no small molecules given off as in classic condensation examples.) Condensation reactions normally produce a small by-product molecule such as water. The reaction is explained as a "step-wise" reaction.

$$\qquad\qquad\qquad H \qquad\qquad\qquad H$$
$$\qquad\qquad\qquad | \qquad\qquad\qquad |$$
$$\qquad\qquad\qquad N \quad O \quad O \quad N$$
$$2\,O=C=N-R \qquad C \qquad R_2 \qquad C \qquad R-N=C=O \;+\; HO\cdot R_3\cdot OH$$
$$\qquad\qquad\qquad\qquad || \qquad\qquad ||$$
$$\qquad\qquad\qquad\qquad O \qquad\qquad O$$
$$\qquad\qquad\qquad\qquad\qquad\qquad\qquad\qquad\qquad\qquad\qquad \longrightarrow$$

$$H \qquad\qquad\qquad H \qquad\qquad\qquad\qquad\qquad H \qquad\qquad\qquad H$$
$$|\qquad\qquad\qquad | \qquad\qquad H \qquad\qquad H \qquad\qquad | \qquad\qquad\qquad |$$
$$N \quad O \quad O \quad N \quad | \qquad O \quad | \qquad O \quad N \quad O \quad O \quad N$$
$$-R \quad C \quad R_2 \quad C \quad R-N-C \quad R_3\cdot N-C \quad R \quad C \quad R_2 \quad C \quad R-NCO$$
$$\quad || \qquad\quad || \qquad\quad || \qquad\qquad || \qquad\quad || \qquad\quad ||$$
$$\quad O \qquad\quad O \qquad\quad O \qquad\qquad O \qquad\quad O \qquad\quad O$$

The ratios of the reactants are important as they influence the properties of the final product. If the ratio of the diol and diisocyanate is one to one, a linear thermoplastic is formed.

When the mole reaction ratio of the diol to diisocyanate is kept in the range of one mole diol to 1.6 to 2.25 moles of the diisocyanate, a viscous (pre)polymer is formed. If care is taken during preparation to keep the isocyanate in excess during the reaction, the prepolymer will have terminal isocyanate groups. The overall molecular weight must be increased by joining prepolymer chains together using either diols or diamines.

When the prepolymer is chain extended with a diol, the polymer formed has only urethane linkages and is strictly a polyurethane. The polymer formed with the diamine chain extender is a poly(urethane urea). The first urethane component is from the initial chain extension when the prepolymer is prepared. A diamine curative will form urea linkages between chains. It is in normal course still referred to as a polyurethane.

$$2\,O{=}C{=}N{-}R \overset{\overset{\displaystyle H}{\underset{\displaystyle N}{|}}}{\;} \overset{\overset{\displaystyle H}{\underset{\displaystyle N}{|}}}{\underset{\underset{\displaystyle O}{\|}}{C}} \; R_2 \; \overset{\overset{\displaystyle H}{\underset{\displaystyle N}{|}}}{\underset{\underset{\displaystyle O}{\|}}{C}} \overset{\overset{\displaystyle H}{\underset{\displaystyle N}{|}}}{\;} R{-}N{=}C{=}O \;+\; NH_2 \cdot R_3 \cdot NH_2$$

$$\longrightarrow$$

$$-R \; \underset{\underset{O}{\|}}{C} \; R_2 \; \underset{\underset{O}{\|}}{C} \; R{-}N{-}\underset{\underset{O}{\|}}{C} \; R_3 \; \underset{\underset{O}{\|}}{C} \; R \; \underset{\underset{O}{\|}}{C} \; R_2 \; \underset{\underset{O}{\|}}{C} \; R{-}NCO$$

2.1.3   Side Reactions

Subsidiary chemical reactions can take place. The major ones are the formation of an allophanate cross-link:

$$\overset{H}{\underset{\underset{O}{\|}}{-N-C-O}} \;+\; -N{=}C{:}O^+ - \quad\xrightarrow{\;100\text{—}120\,°C\;}\quad \overset{H}{\underset{\underset{O}{\|}}{-N-C}}-N-\underset{\underset{O}{\|}}{C}-O-$$

Allophanate link

This reaction normally needs a temperature of between 120 and 140 °C to take place.

A urea group present at 100 °C can react with the isocyanate group to form a biuret linkage:

$$\overset{H}{\underset{\underset{O}{\|}}{-N-C}}-N-H \;+\; -N{=}C{:}O^+ - \quad\xrightarrow{\;>100\,°C\;}\quad \overset{H}{\underset{\underset{O}{\|}}{-N-C}}-N-\underset{\underset{O}{\|}}{C}-N-H$$

Biuret cross-link

2.1.4   Polyureas

If the hydroxyl group (OH) in a conventional diol is replaced by an amine group (NH_2), polyamines are formed. The polyamines can be used in a similar manner as polyols to form a polyurea on reaction with diisocyanate. These polymers do not have any urethane groups, only urea groups. Due to their high curing speed, spraying is the best method of application. They are mainly used in thin coatings. Polyureas are used in the mining industry in wear applications.

$$NH_2 \left[\diagup\diagdown\diagup\diagdown \right]_n NH_2 \quad + \quad O=C=N-R-N=C=O \longrightarrow$$

$$\text{Polyamine} \qquad\qquad\qquad \text{Diisocyanate}$$

$$O=C=N-R-\overset{H}{\underset{}{N}}-\overset{}{\underset{O}{C}}-NH\left[\diagup\diagdown\diagup\diagdown\right]_n \overset{H}{\underset{}{N}}-\overset{}{\underset{O}{C}}-\overset{H}{\underset{}{N}}-R-N=C=O$$

$$\text{Polyamine prepolymer}$$

Further chain extension is carried out using diamine curatives such as DETDA or DMDTA. These diamines are used in both the polyurethane and epoxy industries.

2.1.5 Water Reactions

When a polyurethane prepolymer is reacted with a substance containing a single hydroxyl group such as ethanol or water, reactions will take place that give off carbon dioxide gas. In thin sections the bubbles may be able to escape or be held by inert fillers. They can also be trapped by pressure molding the part.

The major problem is the moisture that is absorbed into the polyurethane system or into the curative and auxiliary materials. Free water will liberate carbon dioxide when the chain extension is carried out. It is important to keep the reactants dry as any moisture that may have come in contact with the prepolymer will react to give an amine and carbon dioxide. This amine reacts with more isocyanates to form a disubstituted diamine. The reaction is outlined in the following equations:

$$R-N=C=O + H_2O \longrightarrow \left[R-NH-\overset{O}{\overset{\|}{C}}-OH \right]$$

$$\longrightarrow R-NH_2 \longrightarrow CO_2$$

$$R-N=C=O + R_1NH_2 \longrightarrow R-\overset{N}{\underset{\overset{\|}{O}}{N}}-\overset{}{C}-\overset{H}{\underset{}{N}}-R_1 \text{ (A very fast reac-}$$

tion)

$$2\,R-N=C=O + H_2O \longrightarrow R-\overset{N}{\underset{\overset{\|}{O}}{N}}-\overset{}{C}-\overset{H}{\underset{}{N}}-R + CO_2$$

Two moles of isocyanate are used per one mole of water. The carbon dioxide released will form bubbles throughout the cast product, forming a spongy instead of solid material, and thus detracting from the properties of the polyurethane. The physical properties will be lowered both from the chemical balance being incorrect and the presence of the bubbles, which prevents the full strength from being developed.

2.2 Raw Materials

TABLE 2.1
Main Polyurethane Ingredients

	Group	Type
1	Polyols	Polyether
		Polyester
2	Diisocyanates	Aromatic
		Aliphatic
3	Chain extender	Diamines
		Hydroxyl (glycols or water)
		Polyols
4	Other chemicals	

2.2.1 Polyols

Flexibility in the polyurethane is provided by the backbone or "soft segment." Polyols provide the soft segment of the polymer and are capped with a hydroxyl group. Unless there are special requirements the polyols are linear (i.e., no branching) and of molecular weight between 400 and 7000. The overall molecular weight of the soft segment controls the frequency of the hard phase and hence the hardness, resilience and stiffness of the final product. The lower the molecular weight, the higher the occurrence of the hard phase.

There are two main groups of polyols used to make castable polyurethanes:

1. Polyethers

2. Polyesters

Polyethers

Polyether diols form a very important segment of the diols used in the manufacture of polyurethanes. The normal route is by addition polymerization of the appropriate monomeric epoxide. The most important polyethers are polypropylene glycol (C3) and polytetramethylene glycol (C4).

$$\underset{\text{Propylene oxide}}{CH_2-\overset{\displaystyle O}{\overset{\diagup\diagdown}{CH}}-CH_3} \quad \xrightarrow{\text{Catalyst}} \quad \underset{\text{Propylene glycol}}{H-O\left[CH_2-\overset{\displaystyle CH_3}{\overset{\displaystyle |}{C}}\right]_n O-H}$$

$$\underset{\substack{\text{Tetrahydrofuran}}}{\overset{\displaystyle H_2 \quad O \quad H_2}{\underset{H_2 \quad\quad H_2}{\overset{C \diagdown \diagup C}{\underset{C-C}{}}}}} \quad \xrightarrow{\text{Lewis Catalyst}} \quad \underset{\substack{\text{Polytetramethylene} \\ \text{glycol}}}{H\left[O \overset{CH_2}{\diagup} \overset{}{\diagdown}_{CH_2} \overset{CH_2}{\diagup} \overset{}{\diagdown}_{CH_2}\right]_n O-H}$$

The polyether glycols produce polyurethanes that are not as strong and tough as the polyester-based polyurethanes but they have far superior hydrolytic stability.

The standard polyol in this group is polytetramethylene glycol (PTMEG), which gives compounds with physical and mechanical properties superior to those produced with polypropylene glycol (PPG)-based systems. The PTMEG produces polyurethanes with excellent mechanical properties and very low abrasion loss.

Polyurethanes based on polypropylene glycol (PPG) do not have as good mechanical and wear properties as the PTMEG-based materials.

Improvements in the performance of the PPG material were made by end-capping the propylene glycol chains with ethylene oxide. The modified PPG gave better processing and performance.

Newer polyethers have been introduced into the market. While based on polypropylene glycol they have superior properties, especially in the high molecular weight range. The Acclaim®[1] polyethers are made using an alkoxylation catalyst that yields a material with very much higher diol content [6]. When high-molecular- weight PPG is made using potassium hydroxide catalyst, there are a large number of chains that only have one hydroxyl group (monol) instead of two. This will severely limit later chain extension. The monol content in a 2000 MW diol is reduced from 6 to 1 mole percent using the new process, while with a 4000 MW diol the monol content is reduced from 33 to 2 mole percent.

The purchase cost of the polyols is as follows:

Classical PPG < EO tipped PPG < Low monol PPG< PTMEG

Polyesters

The chemical structure of the prepolymer influences its chemical resistance. Polyesters, due to their structure, have inherently better oil resistance but

---

[1] Bayer Material Sciences.

lower hydrolytic stability. The ether groups in the polyether urethanes provide better hydrolytic stability and are more flexible.

Polyesters are of three different types:

1. Dibasic acid reacted with diol
2. Polycaprolactone materials
3. Polycarbonate-based materials

Dibasic Acid Type

The classic polyester is made by the reaction of a dibasic acid and a diol with the formation of a polyester and water. The water must be removed.

The general reaction equation is shown below:

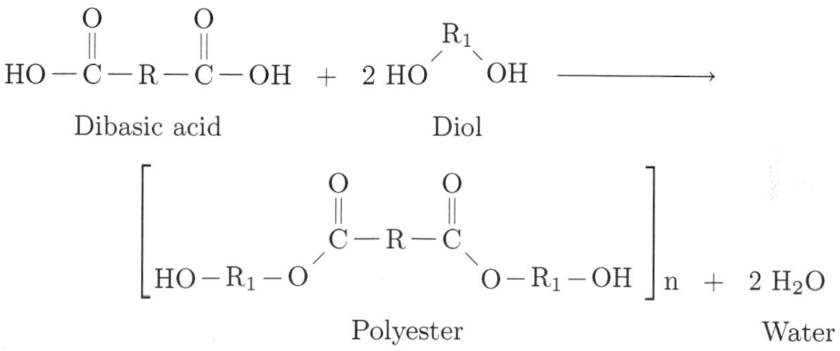

$$HO-\underset{\substack{\|\\O}}{C}-R-\underset{\substack{\|\\O}}{C}-OH \quad + \quad 2\ HO^{R_1}OH \longrightarrow$$

Dibasic acid   Diol

$$\left[HO-R_1-O^{\underset{\substack{\|\\O}}{C}-R-\underset{\substack{\|\\O}}{C}}O-R_1-OH \right]_n \quad + \quad 2\ H_2O$$

Polyester   Water

Polyesters produce strong, tough, oil-resistant materials. The major downside is a lack of hydrolysis resistance. The basic polyesters are prepared by the reaction of a dibasic acid (usually adipic, sebacic or phthalic acid) with a diol such as ethylene glycol, 1,2-propylene glycol, or diethylene glycol. The polyesterification conditions must be such that only hydroxyl groups form the terminal groupings. The water of condensation formed must be removed to a level of 0.03 percent to produce good polyurethanes. Polyurethanes made from these ingredients suffer from relatively poor hydrolytic stability. The reaction between ethylene glycol and adipic acid is shown below:

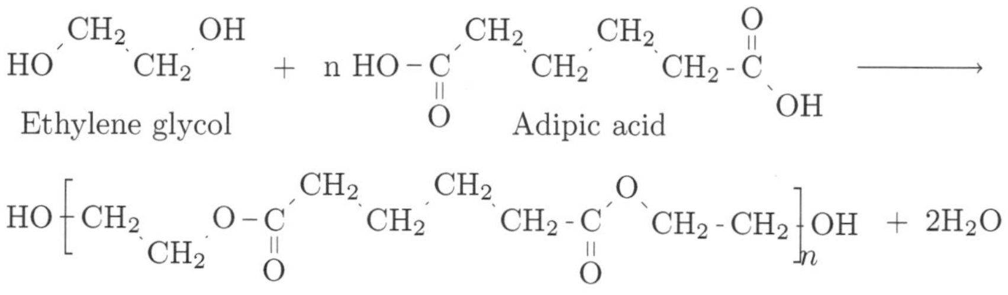

$$HO^{CH_2}CH_2^{OH} \quad + \quad n\ HO-\underset{\substack{\|\\O}}{C}^{CH_2}CH_2^{CH_2}CH_2-\underset{\substack{\|\\O}}{C}OH \longrightarrow$$

Ethylene glycol   Adipic acid

$$HO\left[CH_2^{CH_2}O-\underset{\substack{\|\\O}}{C}^{CH_2}CH_2^{CH_2}CH_2-\underset{\substack{\|\\O}}{C}CH_2-CH_2\right]_n OH \quad + \quad 2H_2O$$

Polyethylene adipate glycol

They offer excellent heat and solvent resistance with high tear and sliding abrasion resistance. Their main drawback is poor hydrolytic stability and susceptibility to fungal attack.

Polycaprolactone

This group of polyesters is made by opening of the caprolactam ring. Caprolactam is also used in the production of nylon. Their structure appears to
provide a degree of protection from hydrolytic attack. They are formed by
the reaction shown below. This method is discussed by Barbier-Baudry and
Brachais [1]. Their hydrolytic properties fall between those of PTMEG and
other polyesters.

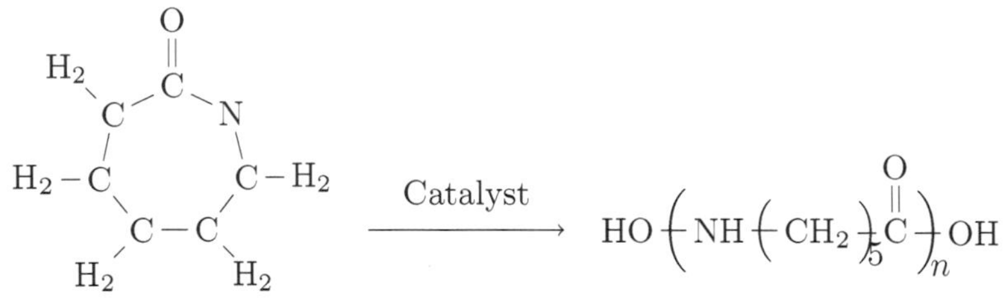

Polycarbonate

When reacting either ethylene carbonate or propylene carbonate with an
aliphatic diamine, a polyurethane can be produced:

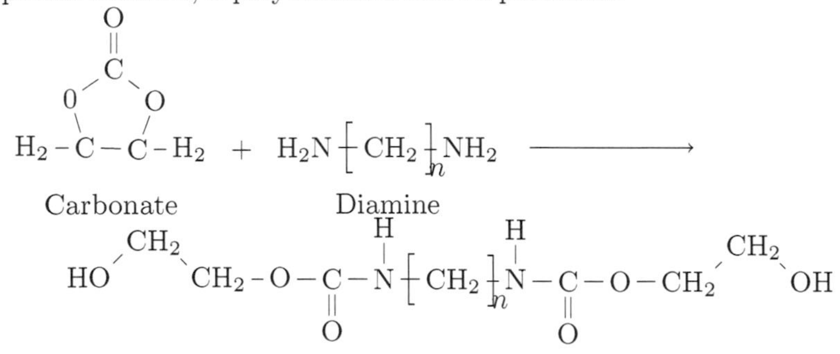

Polycarbonate diol

Poly(ethylene ether carbonate) diols [4], when made into polyurethanes
using MDI and BDO, produce elastomers that have polyester polyol features.
This was shown using ^{13}C NMR. The structure gives rise to the potential for
a very high virtual cross-linking density.

These carbonate-derived polyesters have superior hydrolysis resistance
compared to traditional materials.

2.2.2   Diisocyanates

The isocyanates form the major part of the hard or rigid phase of the
polyurethane. The three main isocyanates used in industry for castable materials are toluene diisocyanate (TDI), 4,4'- diphenylmethane diisocyanate (MDI)
and 1,5-naphthalene diisocyanate (NDI). Aliphatic diisocyanates form a small
segment of the diisocyanate market.

Aliphatics have non-yellowing properties compared to the aromatic diisocyanates.

Aromatic Diisocyanates

In the production of polyurethane elastomers, only diisocyanates are of any use. The major diisocyanates manufactured and used are the 2,4- and 2,6-toluene diisocyanates (TDI) and 4,4′-diphenylmethane diisocyanates (MDI).

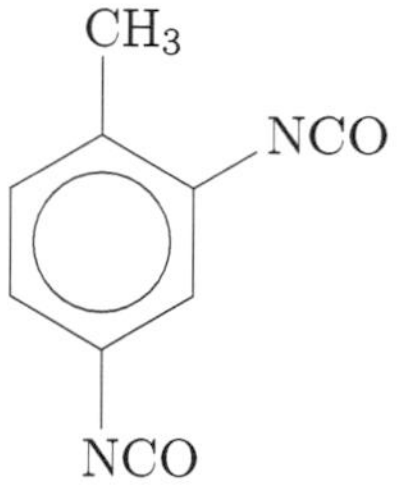

CH$_3$ CH$_3$

NCO   NCO   NCO

NCO

2,4-TDI 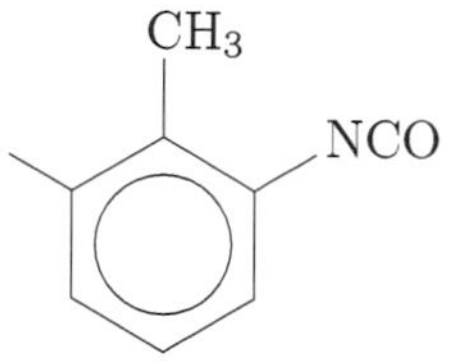   2,6-TDI

$$O=C=N-\bigcirc-CH_2-\bigcirc-N=C=O$$

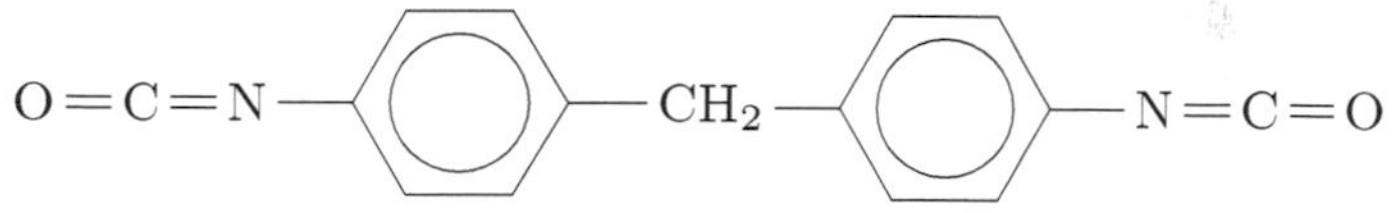

4,4′-MDI

The reactivity of the various isocyanates is important in the processing of any system:

NDI > MDI > TDI

Numbering of Chemical Compounds

The position of groups around an aromatic ring is important in determing the compound's characteristics. Position 1 is always given to the first substitution on the ring:

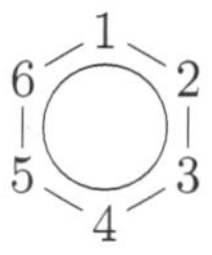

Benzene ring 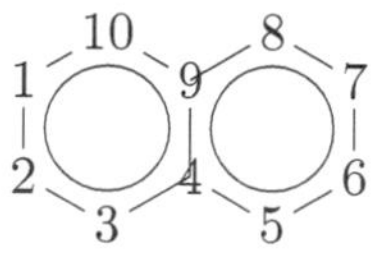   Naphthalene

Organic chemistry textbooks detail more rules for all types of molecules.

The aromatic diisocyanates are generally more reactive than the aliphatic diisocyanates. The position of the isocyanate group relative to surrounding groups controls the reactivity.

The velocity constants of 1,5-naphthalene diisocyanate, where both NCO groups are equal on the ring, are the same. The velocity constants of 2,4-TDI are quoted by Saunders and Frisch[8] as being $k_1 = 42.5$ and $k_2 = 1.6$. The 2,6-isomer of TDI has velocity constants of $k_1 = 5$ and $k_2 = 2$. This indicates that the initial reaction will be fastest with the 2,4-isomer at the 2 position. See insert box for position numberings.

TDI is used either as an 80:20 mixture of the 2,4- and 2,6-isomers or as 100% 2,4-toluene diisocyanate. The current tendency is to use the 80:20 mixed isomers for normal work and the 100% 2,4-isomer TDI for high-performance material. The isomers are separated from crude TDI, which is mainly a 66:33% mixture of isomers with some other combinations.

Pure MDI melts at 38 °C. The MDI tends to dimerize readily so it must be stored at −4 °C and melted just prior to use. The pure MDI can be modified so that it is liquefied at room temperature. The physical properties of the polyurethanes made from these modified isocyanates are inferior to the material made from pure MDI. The reversibility of the reaction is shown in the reaction below [2]:

where R =

Aliphatic Diisocyanates

Aliphatic diisocyanates belong to a different organic group. They do not contain the benzene (aromatic) ring. They may have a ring or a straight-chain structure. See Appendix E for the formulas of typical aliphatic isocyanates. The difference in structure between the aromatic MDI and the aliphatic H_{12}MDI is that the rings in the cyclic portions of the aliphatic material have more hydrogen atoms.

Due to their very much higher cost, aliphatic diisocyanates find use mainly in specialized areas where their special properties, such as non-yellowing in light, are of great importance. The non-yellowing is due to the aliphatic structure of the isocyanate. There is no series of double bonds that cause the yellowing.

The reactivity of aliphatic diisocyanates is low in comparison to aromatic isocyanates. It is a problem in the manufacturing stage when using the prepolymer route. Quasiprepolymers and one-shot reactions require the correct choice of curative and catalyst for the system to work.

The reactivity of aliphatic diisocyanates is low in comparison to aromatic isocyanates. It is a problem in the manufacturing stage when using the prepolymer route. Quasiprepolymers and one-shot reactions require the correct choice of curative and catalyst for the system to work.

Diisocyanates such as p-TMXDI do not form allophanates. This adds to their very stable storage life even at slightly elevated temperatures. The prepolymer has a shorter pot life and gel time compared to an equivalent prepolymer made with the aromatic MDI.

Aliphatic polyurethanes with their non-yellowing also provide sufficient impact resistance so that they can be used in face shields (US Patent 3,866,242). Even with the non-yellowing properties, UV protection agents are often added to materials.

Other commercially available aliphatic and aromatic diisocyanates used in the polyurethane industry are listed in Table 2.2. These diisocyanates represent 3 to 4 percent of the market use. The remainder is TDI and MDI.

TABLE 2.2
Isocyanates Used in Lesser Quantities

Name	Abbreviation
4,4′-Dicyclohexylmethane diisocyanate	$H_{12}MDI$
1,5-Naphthalene diisocyanate	NDI
1,6-Hexamethylene diisocyanate	HDI
Xylene diisocyanate	XDI
Isophorone diisocyanate	IPDI
3-Isocyanatomethyl-3,5,5-trimethylcyclohexyl isocyanate	TMDI

2.2.3 Chain Extenders

The two main groups used as chain extenders are diamines and hydroxyl compounds. Triols are also used where some cross-linking is required. The choice of chain extender depends on the properties required and the process conditions. Diols are the most commonly used hydroxyl compound. In the normal reaction course, diols provide good properties and processing speed with MDI-based prepolymers and diamines with TDI terminated prepolymers.

The molecular shapes of the isocyanate and extender molecules are often considered to play a part in the ease of formation of hydrogen bonding. The molecules must be able to come in close proximity to each other for hydrogen bonding to take place. There must be no steric hindrance to the two chains. Molecules with an even number of carbons are said to allow the hydrogen donor group (NH) to fit more easily to each electron donor group (C=O). When they are odd, the fit is poor, and many groups cannot participate in

TABLE 2.3
Effect of Chain Length on Melting Point

Name	Melting point
$O=C=N$ —CH$_2$—CH$_2$—CH$_2$—CH$_2$—N$=$C$=$O	190 °C
1,4 diisocyanatobutane	
$O=C=N$ —CH$_2$—CH$_2$—CH$_2$—CH$_2$—CH$_2$—N$=$C$=$O	156 °C
1,5 diisocyanatopentane	
$O=C=N$ —CH$_2$—CH$_2$—CH$_2$—CH$_2$—CH$_2$—CH$_2$—N$=$C$=$O	180 °C
1,6 diisocyanatohexane	
$O=C=N$ —CH$_2$—CH$_2$—CH$_2$—CH$_2$—CH$_2$—CH$_2$—CH$_2$—N$=$C$=$O	152 °C
1,7 diisocyanatoheptane	

the hydrogen bonding. Experimentally it has been shown [12] that the melting points of polyurethanes made with a series of aliphatic diisocyanates with different number of carbons in the chain, varied with the number of carbons in the diisocyanate. Those made with an odd number of carbons in the isocyanate had a lower melting point than those on either side with an even number of carbon atoms in the isocyanate chain. This is shown in Table 2.3.

Diamines

Aromatic diamines are the most commercially used chain extenders with TDI based polyurethanes. The rate of reaction of a simple aromatic diamine is too great for normal use. Thus, the rate of reaction is commonly controlled by having substitutes on the aromatic ring. An example is the simple diisocyanate MDA and MOCA with the chlorine atoms in MOCA slowing the reaction to a usable rate.

A chlorine or methyl group ortho (positions 1,2 = ortho) to the amine group gives optimum properties in a diamine-cured compound. It has been found that a methoxy group next to the aromatic amine gives polyurethane with inferior properties.

There are a number of different polyurethane curatives available based on methylene dianiline with different groups next to the amine group (NH$_2$). MOCA with the chlorine atoms has been the most successful.

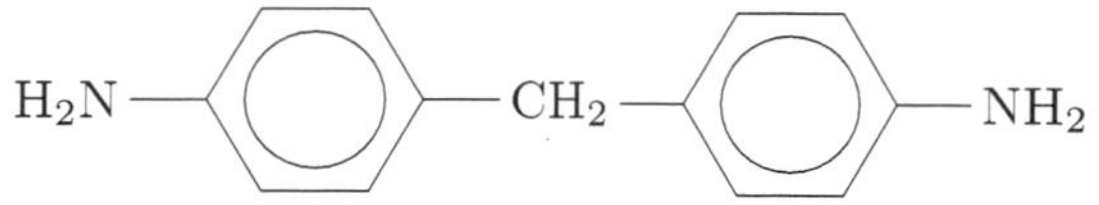

$$H_2N - \bigcirc - CH_2 - \bigcirc - NH_2$$

4,4′-Methylenedianiline (MDA)

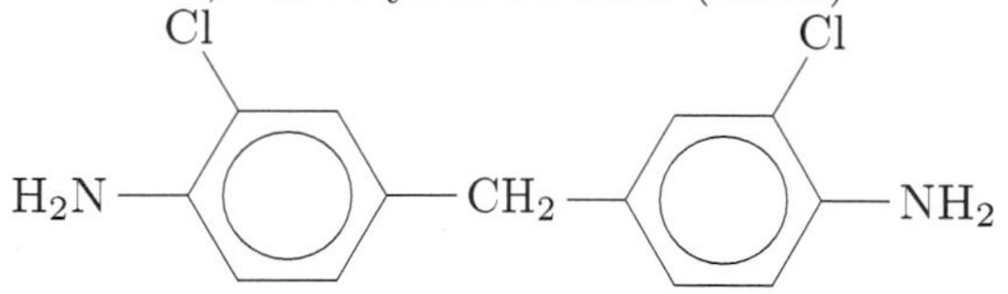

$$H_2N - \bigcirc - CH_2 - \bigcirc - NH_2$$

4,4′-Methylene 2-bis(2-chloroaniline) (MOCA)

Some aromatic diamines used as curing agents, such as 4,4′-diaminodiphenyl derivatives or 3,3′-dichloro 4,4′-diaminodiphenylmethane (MOCA), have been investigated and have been declared to have suspect carcinogenic properties. **All local rules and regulations must be adhered to if these materials are used.**

The mixed isomers of di(methylthio)-toluene diamine are sold under the trade name Ethacure 300 The advantages of Ethacure 300 are that it is a liquid at ambient temperatures and does not have the suspect carcinogenic properties of MOCA.

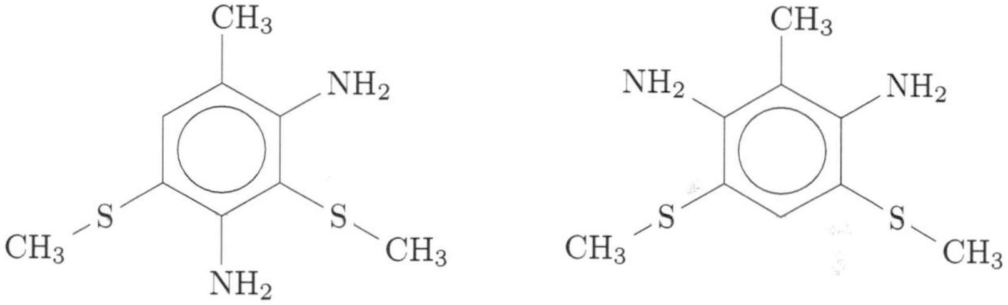

Di(methylthio)-
2,4-toluene diamine

Di(methylthio)-
2,6-toluene diamine

Ethacure 300

Ethacure 300 being a liquid at room temperature, has led to its favoritism in the marketplace for normal production runs. MOCA and Ethacure have advantages over each other in different aspects. In certain mechanical properties Ethacure is superior and in other aspects MOCA is superior. Individual testing needs to be carried out to obtain the best curative for the situation. A newer diisocyanate, Lonzacure®[2] M-CDEA, gives properties superior to Ethacure or MOCA. The processing disadvantage is that it is a solid at room temperature. The melting point of Lonzacure M-CDEA is 88 to 90 °C, which is slightly lower than that of MOCA. The structure of Lonzacure M-CDEA is given below:

---

[2]Lonza Ltd., Basel, Switzerland.

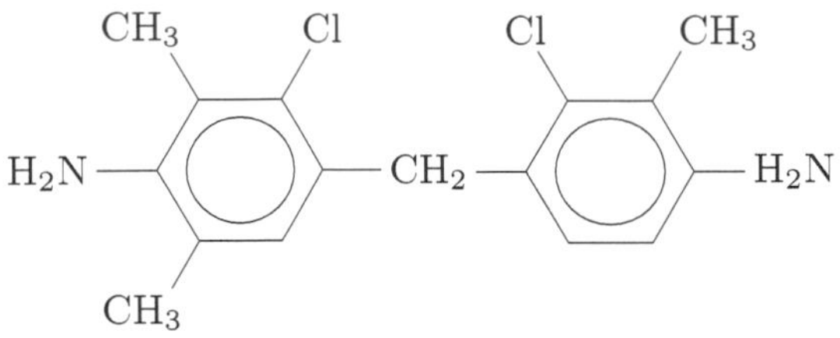

$$CH_3 \quad\quad Cl \quad\quad\quad Cl \quad\quad CH_3$$

$$H_2N - \bigcirc - CH_2 - \bigcirc - H_2N$$

$$CH_3 \quad\quad\quad\quad\quad\quad\quad$$

M-CDEA

Amines produce polyurethanes with better mechanical properties than when diols are used for curing. Amines produce polyurethanes with a lower temperature resistance than when diols are used.

The use of catalysts has been found to direct the cross-linking reactions away from the biuret to the allophanate reactions.

Hydroxyl Compounds

Monols

The simplest hydroxyl compound that can be used is water. The two main disadvantages of water is the fact that carbon dioxide is liberated during the reaction, and the strength of the polymer formed is not very high. Polyurethanes also react with alcohols in a similar manner.

Cast polyurethanes cured with water would need the bulk of the carbon dioxide to be flashed off and the remainder held in place by compression molding.

Low-Molecular-Weight Diols

The chemistry is slightly different when diols are used. The bonds formed are urethane bonds. This means that the final product formed is pure polyurethane. No carbon dioxide is liberated during the reaction. The general reaction was previously described in the chemical reaction section.

The shorter the linear diol chains, the better the compression set and the higher the melt temperature of the polyurethane elastomer. The hysteresis curve shows the least retained energy, thus giving a lower heat buildup under load. These desirable properties can be achieved more readily by the diols ranked in the following series:

Ethylene glycol > 1,3-Propane diol > 1,4-Butane diol >
1,5-Pentane diol > 1,6-Hexane diol.

The most commonly used diol is 1,4-Butane diol. It is generally referred to as just BDO or butanediol. The molecule is shown below:

$$HO-CH_2-CH_2-CH_2-CH_2-OH$$

Butane-1,4-diol (BDO)

Molecular weight = 90.1
Equivalent weight = 45.05

However, one of the main points to be considered in the making of a polyurethane elastomer (apart from the reactivities of the isocyanate and curative) is the water absorption. Special precautions must be taken to keep a diol curative (such as butanediol) sufficiently dry for use. Diols must be kept dry with molecular sieves and dry nitrogen blanketing.

The solubility of the diols in the system must be carefully controlled so that phase separation does not occur. This is of great importance in quasiprepolymers.

Triols

Triols such as trimethylol propane (TMP) provide sites for cross-linking to occur.

$$H_3C-CH_2-\overset{\overset{\displaystyle CH_2-OH}{|}}{\underset{\underset{\displaystyle OH-CH_2}{|}}{C}}-CH_2-OH$$

Trimethylol propane

Triols cause chemical cross-linking that is not dependent on the reaction temperature, as is the case with allophanate and biuret reactions. This form of cross-linking leads to a lower density of the hard phase and a softening of the material. These changes are due to chain separation and a reduction in the hydrogen bonding. Other triols such as glycerol can be used. Glycerol is hygroscopic and is normally supplied with a water content of 2 percent. This causes problems in processing.

Polyols

Linear polyols of various molecular weights can be used in the chain extension of prepolymers. Their main use is to adjust the hardness of the final compound. They act as a reactive plasticizer, and their functionality and molecular weight must be taken into account in the curative calculations. Either polyether or polyester polyols can be used. The most important point is that they must be dry.

2.2.4   Other Chemicals

There are a group of chemicals used in polyurethanes that have an influence
on the physical properties of the final product but do not take part in any
chemical reaction.

Fillers

Fillers are not normally used in polyurethanes to bulk out the product as
they reduce the properties too drastically. This is in contrast to conventional
elastomers where they can be used to reinforce the product. Ultra-fine silica
is used as a thixotropic filler in trowellable polyurethanes.

Plasticizers

Plasticizers can either be reactive or nonreactive. The nonreactive plasticizers
are from the phthalate and ester groups. Typical plasticizers are DIOP, TCP
and Benzoflex 9-88. The level of plasticizer must be controlled because the
physical properties decrease as the levels increase.

Moisture Scavengers

Molecular sieves are synthetic crystalline metal alumino-silicates that have
very fine holes in their structure. The size of the holes is approximately 4 to
5 Å(angstroms) (10^{-10} m) in diameter. The water molecules enter the holes
and are trapped in them. Commercially they are supplied in the form of a
castor oil paste.

Ultraviolet Absorbers

Ultraviolet absorbers can be added to polyurethanes to decrease the yellowing
of the material. They are best used in the aliphatic materials to stop minor
yellowing. In polyurethanes with aromatic diisocyanates, the attack is very
fundamental and hard to stop.

Catalysts

Catalysts speed up certain reactions in the chain extension of polyurethanes.
The catalysts used are those made by specialty suppliers for the polymer
industry and include a range of amines and metal salts.

Nanoparticles

Nanoparticles are now being investigated and used to provide some extra
stiffness to polyurethanes. These are very small, flat, clay-like platelets of
micron size. Montmorillonite of size 2 to 13 microns has been used. They can
have functional groups such as amines attached to them. The particles must

be fully defoliated and wetted by the polyurethane. The functional groups will bond to the hard segments.

2.3 Prepolymers

2.3.1 Introduction

Hepburn [5] gives two main routes for the preparation of polyurethanes based on the relative activities of the various ingredients:

1. Prepolymers variants

   (a) Unstable prepolymers
   (b) Stable prepolymers
   (c) Chain extension/cross-linking to complete production of polyurethane

2. One-shot systems

2.3.2 Prepolymer Variants

Unstable Prepolymers

The unstable prepolymers are typified by the Bayer Vulkalon polyurethane. Vulkalon is generally made by reacting naphthalene diisocyanate (NDI) and hydroxyl-terminated polyesters to form a prepolymer. Traditionally the reactivity of the prepolymer is such that it must be chain extended with a diol or water within a time span of one to two hours. After this time period the prepolymer will react with itself to form an unsatisfactory product. Bayer now supplies into certain markets a product with a shelf life of up to six months.

The best physical properties are obtained with polyethylene adipate as the backbone. After the manufacture of the prepolymer with the polyethylene adipate and NDI, the polyurethane was completed with the addition of BDO. The use of diamines gave too fast a reaction for successful processing. Water and glycols were also used for cross-linking. Glycols do not give the CO_2 gas that the water chain extension does. To improve the low temperature properties a different backbone needs to be used. A mixture of polyethylene and polypropylene adipates has been found to be suitable. The physical properties are not as good as with the single polyol.

Stable Prepolymers

Following the early developments using NDI, it was found that by using TDI instead, a far more stable prepolymer could be made. Stable prepolymers are normally made using either polyesters or polyethers that have been reacted with a slight excess of a diisocyanate such as toluene diisocyanate (TDI) or methylene diisocyanate (MDI). Providing the storage is moisture-free, the stable prepolymer may be kept for months before use. The polyurethane is prepared by chain extension with diols or diamines.The starting point is to react two moles of the diisocyanate with one mole of the polyol. The ratios of the two ingredients can be varied to change the overall properties.

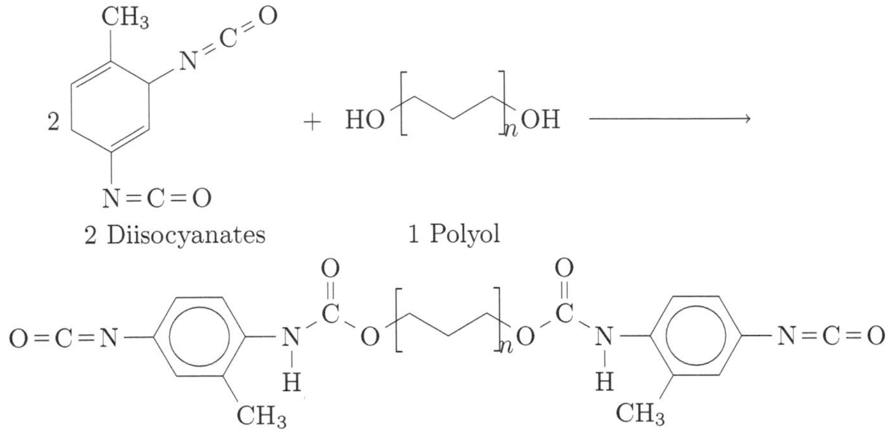

2 Diisocyanates   1 Polyol

Prepolymer

Prepolymer Length

The basic design is to have two molecules of diisocyanate and one of polyol. If the reaction continues at this stage there are several different scenarios:

1. The start chain reacts with a further polyol and diisocyanate

2. Shorter chains combine with a polyol link

    (a) The %NCO is decreased in these cases

3. Shorter chains combine with a polyol link

4. It reacts with a mono polyol or isocyanate

5. There is no more isocyanate

6. The reaction is stopped at a practical limit

The "n" in formulas in diagrams represents the number of repeat units, normally 4 to 25.

The reaction is exothermic so the temperature must be controlled to prevent side reactions.

TABLE 2.4
Effect of Substitution Position on Reaction
Velocity

| | Reaction Velocity | |
	75 °C	100 °C
2- position NCO	0.72	3.2
4- position NCO	3.4	8.5

When the diisocyanate and polyol are reacted together, the diisocyanate is kept in excess as long as possible. The overall conditions are kept slightly acidic to reduce side reactions taking place. The acidity is a must when there is more polyol than diisocyanate present. The older-style PPG polyols may contain traces of potassium hydroxide (KOH). The KOH is a very active alkali. Even small traces have a strong effect.

The most reactive isocyanate group (-NCO) in the structure will react with the hydroxyl (-OH) group first. The reaction will tend to favor the most reactive site. The reaction is allowed to proceed until almost all the diisocyanate is used up. The reaction is stopped before the viscosity increases too much. The temperature and speed of the reactions are such that catalysts are normally not used in these reactions.

The reaction rates at the various positions on the ring vary greatly. For example, with TDI the relative rates are shown in Table 2.4.

The reactions are exothermic so the rate of addition must be controlled to prevent side reactions.

Early work published in Saunders and Frisch, volume II [8], shows how various properties change with different diisocyanate-to-polyol ratios. As the NCO to OH ratio increases from 2 to 2.75, the main physical properties of tensile strength, modulus, tear strength and hardness increase. Other positive improvements are compression set and resilience. The pot life and Taber abrasion properties of the final polyurethane decrease as the ratio increases.

2.3.3 One-Shot Systems

Single-Step

In the single-step system the entire reaction takes place when all three basic components plus catalysts and pigments are mixed together. Careful use of catalysts enables the preparation of the prepolymer and the subsequent chain extension to take place in one step. As all the reactions are exothermic, heat is liberated and must be removed; otherwise part distortion will take place.

The commercial use of this system requires careful metering and temperature control of the various streams. Due to the heat liberated, this system is best suited to long runs of thin wall products.

The systems may be either a three-dimensional network based on covalent bonding or linear extended chains bonding through hydrogen bonds.

Quasiprepolymers

The high exotherm and the number of feeds have led to the quasiprepolymer variation of the one-shot technique becoming popular. Quasiprepolymers are a version of the standard prepolymer method where some of the polyol is reacted with an excess of diisocyanate. This is later reacted with the remainder of the polyols, including the chain extender and some catalysts. The advantage of this scheme is that the exotherm is spread between the preparation of the quasiprepolymer and the final cured part. In MDI systems, BDO is often used as the curative. As BDO's molecular weight is low, the amount added is low. By dissolving it in the polyol, a more even mix ratio can be obtained.

The polyol, the chain extender and any catalysts are mixed and stored as a separate item. The second item, namely the diisocyanate prepolymer, is kept as a "B" part until the material is ready for use. When the two parts are mixed, the chain extension and cross-linking take place simultaneously. Methylene diisocyanate (MDI) and a catalyst such as an organic tin or bismuth salt, provide the required rate of reaction. The size of the exotherm is such that the reaction proceeds to its full extent.

2.4   Urea and Urethane Reactions

2.4.1   Introduction

Compounds that contain the N-H group are all potentially reactive toward isocyanates. Simple primary amine groups are very reactive even at temperatures of between 0 °C and 25 °C, giving disubstituted ureas in high yields. Secondary aromatic amines are slightly less reactive. The basic ratios are given below. Substitution on the aromatic ring is used to control the rate of reaction.

Compounds containing the O-H group such as diols will, under the appropriate conditions, react with isocyanates to give a urethane, or in pure organic literature, a carbamate. Secondary alcohols react at about one third of the rate of primary alcohols. Complex alcohols react very slowly with isocyanates to form a mixture of urethanes and olefins.

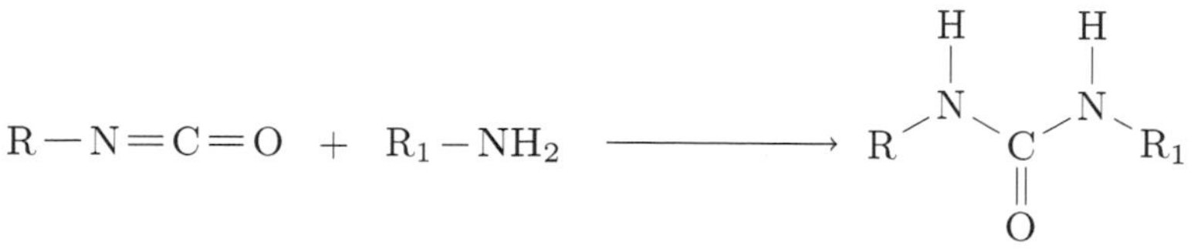

$$R-N=C=O \ + \ R_1-NH_2 \longrightarrow R \underset{\underset{O}{\overset{\|}{C}}}{\overset{\overset{H}{\underset{|}{N}}}{}} \overset{\overset{H}{\underset{|}{N}}}{} R_1$$

Primary amine

Disubstituted urea

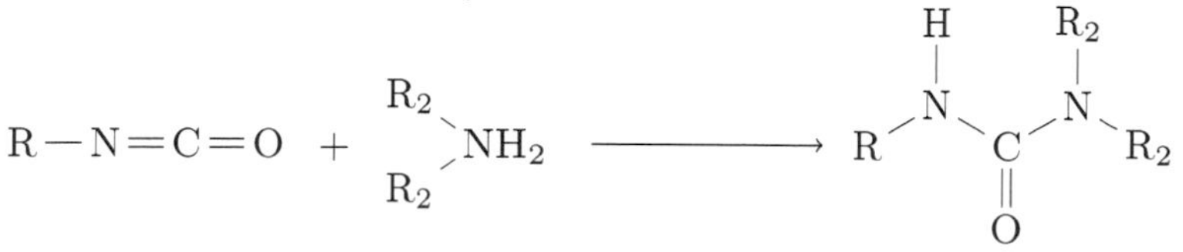

$$R-N=C=O \ + \ \overset{R_2}{\underset{R_2}{}} NH_2 \longrightarrow R \underset{\underset{O}{\overset{\|}{C}}}{\overset{\overset{H}{\underset{|}{N}}}{}} \overset{\overset{R_2}{\underset{|}{N}}}{} R_2$$

Secondary amine

$$R-N=C=O \ + \ R_3-O-H \longrightarrow R \underset{\underset{O}{\overset{\|}{C}}}{\overset{\overset{H}{\underset{|}{N}}}{}} \overset{\overset{R_3}{O}}{}$$

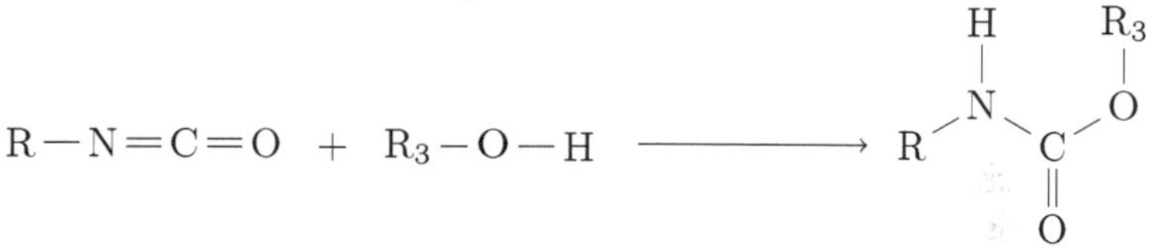

Alcohol

Urethane

2.4.2 Reaction Speed

Isocyanates and amines react together to form ureas. Primary aliphatic amines react very quickly at temperatures down to ambient while secondary aliphatic and primary aromatic amines react less quickly. The reaction rate of secondary aromatic amines is the slowest. The speed of the reaction can further be modified by the addition of substitutes near the amine group.

The control of the speed can either be electronic as illustrated by the effect of the chlorine in the MOCA ring or by stereochemical influences where the groups next to the amine group have a very strong hindrance to the curing. This is illustrated by the effect that the sulfur molecule has on the speed of DMTDA (Ethacure 300) versus that of DETDA (Ethacure 100). The sulfur group slows the speed to a level suitable for polyurethane processing.

The choice of the curative will depend on the polyol–isocyanate system, and the processing temperature used. The normal range of amine curatives tend to give too fast a cure rate when used with MDI-based prepolymers, and give the best results when used with castable systems based on TDI. With TDI-based systems no catalysts are normally needed.

The structure of the aromatic amines and their isomers is shown below:

3,5-Diethyltoluene 2,4-diamine   3,5-Diethyltoluene 2,6-diamine
       2,4-isomer 80%   2,6-isomer 20%
                   DETDA (Ethacure 100)

3,5-(Dimethylthio) 2,4-diamine   E100 2,6-isomer 20%
       2,4-isomer 80%
                   DMTDA (Ethacure 300)

2.5 Chain Extension

In the preparation of a prepolymer, every effort is made to prevent the formation of any unplanned branching such as biuret groups. The prepolymer is essentially linear except when some cross-link sites have been introduced using a multifunctional isocyanate or triol.

2.5.1 Urea

If a diamine curative is used, the amine groups (NH_2) will react with an isocyanate group of the prepolymer to form a urea bond. The remaining amine group will react with a further isocyanate group to extend the chain in the same manner as for hydroxyl extenders.

A diol such as PPG or PTMEG has 5 to 28 repeat units (382—1982 MW). On reaction with a diisocyanate, a urethane bond is formed. A further diol can react with the second isocyanate group, thereby lengthening the chain. Final chain extension (curing) takes place when either a urethane or urea group is added on to form a longer chain. As the reaction proceeds, the mixture first thickens and then gels before becoming initially a brittle solid and then an elastomeric material on final curing. This is an ideal situation but in practice side reactions can occur.

Any variation of the substitutes next to the amine group greatly affects the rate of reaction. A comparison between methylene diamine and MOCA shows the chlorine atoms adjacent to the amines slow the reaction rate right down.

A polyurethane made from a 90 Shore A PTMEG prepolymer and two curatives [3] had excellent physical properties but the pot lives for the two were:

4,4'-Methylene dianiline (MDA)

4,4'-Methylene-bis(2-chloroaniline (MOCA)

2.5.2 Urethane

The use of hydroxyl compounds is normally with either MDI or NDI systems. The simplest hydroxyl compound water can be used. The disadvantage of this is the evolution of carbon dioxide. Water (moisture in air) is normally only used in film-type situations or where the bulk of the gas can be removed.

Where:
Ar - Aromatic or aliphatic group
Ur - Urethane group

Diols are predominately used for chain extensions to give a product with only urea bonds. The progress is similar to the formation of the poly(urea urethane) but with only urethane bonds.

All these reactions are idealized and in reality there are some side reactions taking place. The materials in normal use are of commercial quality and besides there being various isomers, there are a number of impurities. A change in supplier may give variations in the final properties. The reaction rates of both isocyanate groups, when using 4,4'-MDI (the normal pure MDI), are the same.

2.5.3   Hydrogen Bonding

The individual chains of polyurethane comsist of a large number of atoms joined together by covalent bonds. These chains consist of long sections of hydrocarbons (from the initial diols) joined by aromatic or aliphatic hard segments that contain urethane or urea groups (from the isocyanate and chain extenders). The long, more flexible sections are called the soft segment and the urethane/urea sections are the hard segment.

These chains are separate except where joined by allophanate, biuret groups or the use of triols in the system. In the hard segment the urethane and urea groups have electrostatic charges on some of the hydrogen, oxygen and nitrogen atoms. These charged atoms form dipoles that attract another atom of opposite charge, thus forming a hydrogen bond [7].

$$
\overset{\delta+}{H} \underset{\underset{O}{\diagdown \delta-\diagup}}{} \overset{\delta+}{H} \qquad \overset{\delta+}{H} \underset{\underset{\underset{\underset{\delta+}{H}}{|}}{N}}{\diagdown \delta-\diagup} \overset{\delta+}{H}
$$

These hydrogen bonds are of lower strength than the covalent bonds in the rest of the chains but are still sufficient to form a strong compound. The hydrogen bond is approximately 5 kcals/mol compared to the covalent bond of about 50—100 kcals/mol.

During the curing and post curing, the molecules line up and the hard segments tend to agglomerate in groups where the hydrogen bonding takes place. The size of these agglomerations is approximately 25 by 55 Angstrom units in size.

See also the illustrations in Szycher's *Handbook of Polyurethane* [10] (Figures 3.1 and 3.2) for further illustrations. It must be remembered that all these reactions are in three dimensions.

$$
\begin{array}{ccccc}
O & O & O & & O \\
\parallel & \parallel & \parallel & & \parallel \\
-O-C-N-R_1-N-C-R_3-N-C-N-R_1-N-C- \\
\quad\; | \quad\;\; | \quad\quad\quad\; | \quad\;\; | \quad\;\; | \\
\quad\; H \quad\; H \quad\quad\quad H \quad H \quad H
\end{array}
$$

$$
\begin{array}{ccccc}
O & O & O & & O \\
\parallel & \parallel & \parallel & & \parallel \\
-O-C-N-R_1-N-C-R_3-N-C-N-R_1-N-C- \\
\quad\; | \quad\;\; | \quad\quad\quad\; | \quad\;\; | \quad\;\; | \\
\quad\; H \quad\; H \quad\quad\quad H \quad H \quad H
\end{array}
$$

Hydrogen bonding between O and H

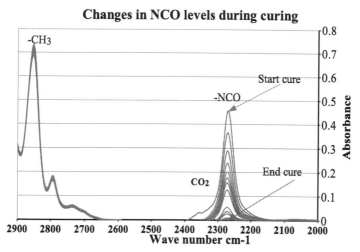

FIGURE 2.1
Decrease in NCO on curing.

Bulky side chains in the hard segment will tend to make the hydrogen bonding more difficult. The bulky side chains of Ethacure 300 will lower the hardness of a material by several points compared to that of a material cured with MOCA. The use of triols or macro diols in the curing phase will have a softening effect. If a trifictional isocyanate or hydroxyl is used in the initial preparation of the prepolymers, this softening does not happen.

Studies in the mid-range infrared spectrum using Fourier Transform Infrared (FTIR) spectroscopy can show the course of the basic curing and the formation of hydrogen bonding. The NCO group has an intense band at the wavenumber 2273 cm^{-1} (note this is near but not at the same number as carbon dioxide). As the final cure progresses, the band will decrease in intensity until it fully disappears. Figure 2.1 illustrates this.

Using a straight polyurethane system, for example a PTMEG/MDI prepolymer, there will initially be an absorption band at approximately 1733 to 1725 cm^{-1} [9]. As the cure starts, this band will initially increase slightly in intensity. Then a second band at 1700 cm^{-1} will appear due to hydrogen bonding. From the size of the split it is deduced that the bulk of the hydrogen bonding is in the hard segment.

$$\diagup C = O \qquad\qquad 1730 \text{ cm}^{-1}$$

$$\diagup C = O\text{-}\text{-}H\text{-}N\diagup \qquad\qquad 1702 \text{ cm}^{-1}$$

Due to the shape of the molecules and the fact that they are twisted and bent, the bonding is in all three planes and not as a flat object.

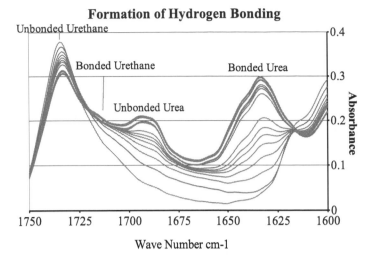

FIGURE 2.2
Formation of hydrogen bonding over time.

When a diamine curative is used, urea bonds are formed. These bonds also play an important part in the hydrogen bonding process. The unbonded urea has an absorption band at approximately 1690 cm^{-1} and the bonded urea at 1682 cm^{-1}. The time sequence spectra (Figure 2.2) show the formation of the bonded urea bands over the first hours of curing.

Different references give variations in the values for the different bands. These vary depending on the calibration of the instrument used, the exact chemistry of the system, and the method of preparation and testing of the sample. Samples tested in a solvent may indicate the carbonyl band at a wavenumber of 1745 cm^{-1}, whereas the solid material will be at approximately 1730 cm^{-1}.

Maximum properties are formed during the post curing as the molecules align themselves and the hydrogen bonding reaches a maximum.

Some studies have shown that the 2,6-TDI isomer in polyurethanes is high hydrogen bonded. The 2,4-isomer however is only 50 percent hydrogen bonded [11].

The ability of the polyurethane chains to form zones with a high concentration of hard segments is vital for the formation of material that has excellent all-round properties. Some of the factors influencing the hydrogen bonding include:

- The use of long-chain diols (macro diols)
- Chemical structure of the curative
- The use of triols

- Plasticizers

Polyurethanes with a very high carboxyl (-C=O) content are also believed to start hydrogen bonding at the ether groups in the soft segment chain.

2.5.4   Three-Dimensional Cross-Linking

Hydrogen bonding takes place in three dimensions. Molecules are not planar but have various atoms and groups protruding at various angles. The hard segments tend to form cylindrical clusters with flexible chains in between.

$$3 \underset{\text{Terminal NCO groups}}{R_3 - Ar - N = C = O} \quad + \quad \underset{\text{Trifuctional hydroxyl}}{HO - \overset{\overset{\displaystyle OH}{\displaystyle |}}{R} - OH} \longrightarrow$$

$$R_3 - Ar - \underset{\overset{|}{H}}{N} - \overset{\overset{\displaystyle O}{\parallel}}{C} - O - R \begin{cases} O - \overset{\overset{\displaystyle O}{\parallel}}{C} \diagdown \underset{\overset{|}{H}}{N} - Ar - R_3 \\[2ex] O - \underset{\overset{\parallel}{O}}{C} \diagdown N - Ar - R_3 \end{cases}$$

Some cross-linking takes place in the form of side reactions such as biuret and allophanate bonds. Deliberate introduction of covalent cross-linking can be carried out by the introduction of multifunctional agents (mainly triols) into either the prepolymer or chain extension system.

Triols

When triols are introduced into the prepolymer, they react in a slightly different manner when used as a curative. Prepolymers made containing some triols have a larger hard segment at the point where the triol reacts with the diisocyanate.

At this point the chains have not fully developed and the hard segments will have joined the chain at random lengths in the structure. This has the effect of increasing the viscosity of the mix as well as improving the compression set and swelling in solvents. This is important in very hard compounds. A trifunctional isocyanate such as Tolonate HDT from Rhone Poulenc will do the same as TMP but with less decrease in dynamic properties.

When the TMP is used as a curative, the reaction speed is slowed and the material is softened due to the disruption of the chains. The bonds will basically only be at the end of the chains.

2.5.5   Catalysts

Catalysts have been classically defined as materials that, when present in relatively small amounts, influence the speed of a chemical reaction while not undergoing any permanent chemical change themselves. A catalyst is generally considered to speed up a reaction. It can also slow it down.

There has been a group of catalysts developed by polyurethane raw material suppliers with the polyurethane market in mind. Mercury-based catalysts, due to their toxicity, have been banned in some countries or severe restrictions placed on their use. Bismuth-based catalysts are recommended in their place.

Catalysts influence different reactions more vigorously than others. The reactions in the polyurethane industry of interest are:

- Reactions with amines

- Reactions with diols OH-NCO

- Reactions with isocyanates and diols NCO-OH

Reactions with Amines

The most commonly used catalysts when working with TDI/amine systems are three acids, namely:

1. Adipic acid

2. Azealic acid

3. Oxalic acid

These acids are normally used at a rate of 0.3 to 0.6 parts per hundred of prepolymer. Adipic acid is particularly effective if MOCA is being used. Diagrams of the acid catalysts are shown below. Amine catalysts such as DABCO® or DABCO 33LV® are also used with amine cures.

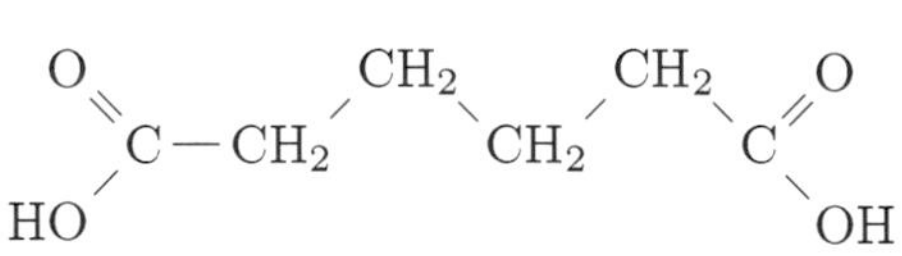

Adipic acid

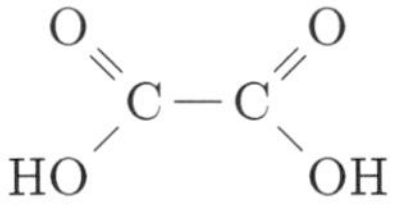

Oxalic acid

$$CH_3-\underset{\underset{OH}{\overset{|}{CH_2}}}{\overset{\overset{CH_2}{\overset{|}{\,}}}{C}}-C\underset{OH}{\overset{O}{<}}$$

Azealic acid

DABCO 33 LV is also known as triethylene diamine. It is a cage-like compound with no steric hindrance, which helps make it a very effective catalyst. It is reactive at close to ambient temperatures. The application range is in the range of 0.3 to 0.6 parts per 100 prepolymer.

1,4-Diazabicyclo[2.2.2]octane
DABCO 33LV

To obtain the desired rate, a mixture of an acid (such as oleic) with a bismuth catalyst can be used. Bismuth catalysts include bismuth neodecanoate and bismuth octoate.

Reactions with Diols

Tin-based salts such as stannous oleate or a dibutyl tin salt (Dibutyl tin dilaurate TDTL), will activate the OH-NCO reactions. This group speeds up the overall reaction rate as well. They also even out the reactivities of low and high molecular weight polyols.

Compared to polyester-based systems polyether diols require a greater degree of activation. The solution is to use a combination of a tin and an amine catalyst to obtain the required speed.

Reactions with Isocyanates and Diols (NCO-OH)

Traditionally mercury catalysts have been the catalyst of choice but because they are highly dangerous, they have, to a very large degree, been replaced by bismuth-based catalysts. The mercury catalysts are both toxic and caustic.

Bismuth catalysts promote the isocyanate and OH reaction from the isocyanate side. This has the effect of reducing the water reaction and hence reduces the liberation of carbon dioxide gas. They are particularly useful in controlling the reactions when used in one-shot systems.

General

Catalysts are commonly used to control the pouring and gel times of polyurethane systems. As catalysts are used in small quantities, the addition must be carried out very carefully, otherwise very erratic results may be obtained. Catalysts may be made into a concentrated solution with an inert carrier such as a plasticizer. This reduces a potential dispensing error. They may also be added to a liquid chain extender.

Tin catalysts are very moisture sensitive and as only small quantities are used at a time, the bulk supply will deteriorate over time with repeated drum openings. The bulk container can be subdivided.

Raw material suppliers have a large range of catalysts available, each with special properties. They should be consulted when required. Some of the catalysts are heat activated and start very slowly and then react faster as the temperature increases.

Catalysts to some degree have an affect on the final properties of the product. The most commonly affected properties are high-temperature heat stability and water resistance. If these are of importance, full evaluations should be carried out.

2.6   Degradation

Degradation occurs under a variety of conditions. Two major causes are:

1. Prepolymer aging
2. Alkaline/ acid hydrolysis

2.6.1   Prepolymers

If the acidity of the prepolymer is not correct or if the material is stored at too high a temperature, the material will start to thicken. The reaction is irreversible. The level of isocyanate also falls under these conditions.

Isocyanate   Urethane   Allophanate

2.6.2   Polyester Elastomers

Polyesters are attacked by both dilute acids and alkalis. At elevated temperatures they are also attacked by water, which is normally either very slightly acid or alkaline.

The method of the alkali attack is as follows:

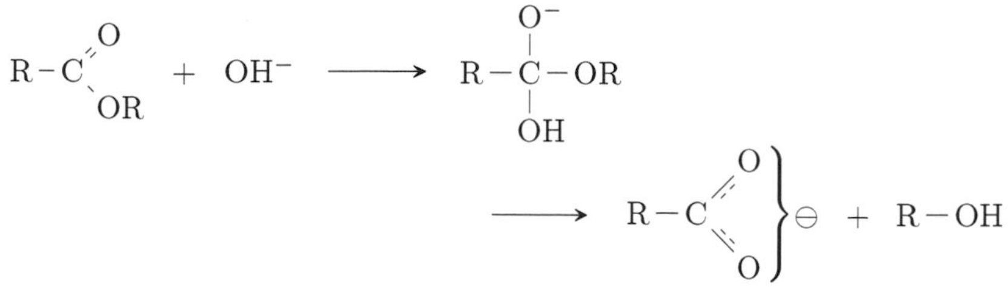

$$R-C\overset{O}{\underset{OR}{\diagdown}} + OH^- \longrightarrow R-\underset{\underset{OH}{|}}{\overset{\overset{O^-}{|}}{C}}-OR$$

$$\longrightarrow R-C\overset{O}{\underset{O}{\diagdown}} \Big\} ^\ominus + R-OH$$

The mode of attack by acids is illustrated below:

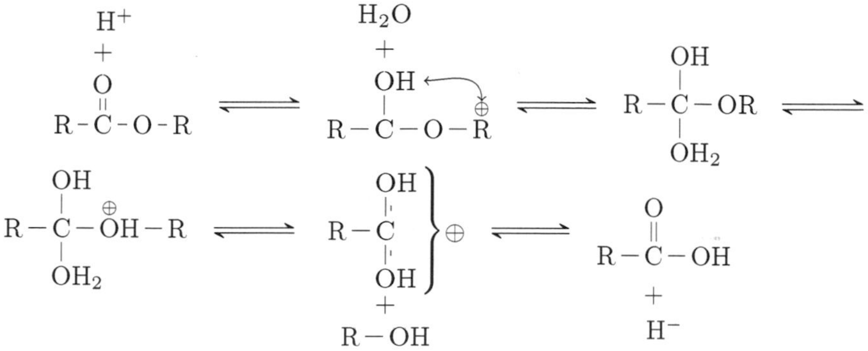

$$\underset{R-\overset{\overset{O}{\|}}{C}-O-R}{\overset{H^+}{\overset{+}{}}} \rightleftharpoons \underset{R-\overset{\overset{OH}{|}}{\underset{|}{C}}-O-\overset{\oplus}{R}}{\overset{H_2O}{\overset{+}{}}} \rightleftharpoons R-\underset{\underset{OH_2}{|}}{\overset{\overset{OH}{|}}{C}}-OR \rightleftharpoons$$

$$R-\underset{\underset{OH_2}{|}}{\overset{\overset{OH}{|}}{C}}-\overset{\oplus}{OH}-R \rightleftharpoons R-\overset{\overset{OH}{|}}{\underset{\underset{OH}{|}}{C}} \Big\} \oplus \rightleftharpoons \underset{\underset{H^-}{+}}{R-\overset{\overset{O}{\|}}{C}-OH}$$

$$\overset{+}{R-OH}$$

Carbodiimides are made from isocyanates under controlled conditions.

$$2\,R-N{=}C{=}O \longrightarrow R-N{=}C{=}N-R + CO_2$$

Isocyanate   Carbodiimide

The reaction of the carbodiimides with the carboxylic acid generated in the hydrolysis is illustrated in the following reaction:

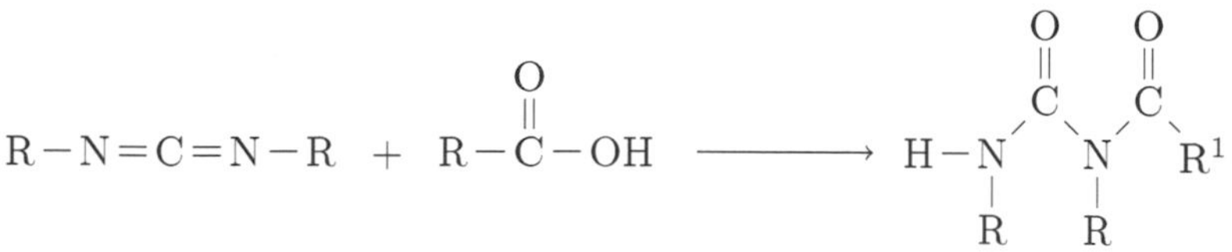

$$R-N{=}C{=}N-R + R-\overset{\overset{O}{\|}}{C}-OH \longrightarrow H-\underset{\underset{R}{|}}{N}\overset{\overset{O}{\|}}{\underset{\diagup}{C}}\underset{\underset{R}{|}}{N}\overset{\overset{O}{\|}}{\underset{\diagdown}{C}}R^1$$

References

[1] D. Barbier-Baudry, L. Brachais, A. Cretu, R. Gattin, A. Loupy, and D. Stuerga. Synthesis of polycaprolactone by microwave irradiation — an interesting route to synthesize this polymer via green chemistry. *Environmental Chemistry Letters*, 1(1):19–23, March 2003.

[2] Dow Chemical Company. Isonate 143L modified MDI. *Trade literature brochure*, pages 1–4, 2001.

[3] DuPont. Methylene dianiline cure — Information sheet. *Trade literature brochure*, pages 1–4, 1973.

[4] R. F. Harris and M. D. Joseph. Polyurethane elastomers based on molecular weight advanced poly(Ethylene ether carbonate) diols. I. Comparison to commercial diols. *Journal of Applied Polymer Science*, 41:487–507, 1990.

[5] C. Hepburn. *Polyurethane Elastomers*. Applied Science Publishers, London, England, 1982.

[6] B. D. Lawrey and N. Barksby. Improved processability and performance in MDI elastomers based on low monol polyols. In *Polyurethanes Expo 2003*, pages 260–267, Orlando, FL, 2003. Polyurethanes Expo 2003.

[7] R. I. Morrison and R. N. Boyd. *Organic Chemistry*. Allyn and Bacon Inc., Boston, 4th edition, 1983.

[8] J. H. Saunders and K. C. Frisch. *Polyurethanes Chemisty and Technology Part One*. Interscience Publishers, New York, 1962.

[9] R. W. Seymour, G. M. Estes, and S. L. Cooper. Infrared studies of segmented polyurethane elastomers. I. Hydrogen bonding. *Macromolecules*, 3(5):579–583, September 1970.

[10] M. Szycher. *Szycher's Handbook of Polyurethane*. CRC Press, Boca Raton, FL, 1st edition, 1999.

[11] H. Ulrich. *Chemistry and Technology of Isocyanates*. J. Wiley & Sons, New York, 1996.

[12] P. Wright and A.P.C. Cumming. *Solid Polyurethane Elastomers*. Maclaren and Sons, London, England, 1st edition, 1969.

Part II

Health and Safety

3

Health and Safety

3.1 Introduction

Socially, economically and legally it is very important to have a workplace that is a safe and healthy environment to work in. If chemicals in polyurethane manufacturing are not used in a safe and careful manner, they can cause harm to workers in the short and long term. There are laws and regulations regarding the use of polyurethane ingredients as well as the associated machinery. The local laws and regulations must be studied, implemented and adhered to when working with polyurethanes and all auxiliary materials.

Warning

It is strongly advised, and in many states it is law, that workers should be medically checked prior to commencement of employment. The isocyanate vapors can bring on lung inflammation problems. They can also over time be a sensitizer.

MOCA has strict safety guidelines that must be obeyed.

3.1.1 Rationale

There are two major chemical areas of health concern in the polyurethane industry. The first is isocyanates in their free form as a liquid or in the vapor form. The second is amines. These can either be curatives or catalysts. One curative MOCA (CAS Number 101-14-4) is classified as a suspect carcinogen in most parts of the world and is subject to many rules and regulations regarding its use. MOCA (MBOCA) is known by many different names, and are listed on websites such as http://www.chemindustry.com/apps/chemicals.

Even traces of isocyanate vapor in the atmosphere can cause bronchial troubles, either in the short or long term. Operators who have asthmatic problems are extremely sensitive to isocyanate vapors and must steer clear of any exposure.

All chemicals used in the polyurethane industry can cause some harm. They must all be treated with both care and respect.

3.1.2   Work Environment

The overall work environment of a polyurethane workshop is very prone to becoming untidy and hazardous due to the nature of the operations being carried out, for example, hot slippery chemicals and hot molds and curing ovens. The layout of the workflow must be under constant review as the emphasis of the product range changes. It is also very important that the safety of the operation is controlled to a large extent by the inclusion of engineering methods of hazards reduction rather than relying mainly on personal protective equipment (PPE).

3.1.3   Acute Exposure

Acute exposure generally refers to single-dose, high-concentration exposures over short periods. Some of the chemicals used in the polyurethane industry can cause acute health problems and have an immediate effect on the health of people exposed to it. The most prominent of the chemicals are the isocyanates. People with bronchial problems can have an immediate attack. It is often suggested that all employees be screened for lung function and for potential problems. Isocyanates also have the potential to sensitize people and they can develop problems in the future.

If mercuric catalysts are used, they must be handled with great care. They can cause burns to the skin that will develop over the next twelve hours subsequent to exposure.

3.1.4   Chronic Exposure

Chronic exposures refer to repeated or continuous exposure over long periods. Amine-based materials such as the amine curatives and catalysts have a potential for problems to occur after a number of years. The highest profile potential problem material is MOCA, which has been the subject of numerous studies. There have been very few cases of fatalities due to exposure to MOCA. The initial casualties also had exposure to precursor materials that were known carcinogens. Further details are very scant.

3.2   Workplace

The workplace environment can be divided into several sections. including:

- Chemicals:
  - Solid
  - Liquid

    − Vapor

- Heated areas
- Finishing
- Engineering

3.2.1   Information Sources

Safety Data Sheets

Safety Data Sheets (SDSs) are replacing Material Safety Data Sheets (MS-DSs) as the prime source of safety data. They are a more globally standardized form.

Safety Data Sheets (SDSs) are a means of communicating their properties and safe-handling procedures to users of industrial and laboratory chemicals. They also give other professionals, such as doctors, paramedics, firefighters and emergency services, information regarding the nature of any problem situations that they may be called upon to assist in.

Initially MSDSs were in many different formats and levels of completeness. They have now been standardized into a recommended format (SDS) so that the information is readily available.

International codes of practice for the preparation of SDSs (e.g., America OSHA, Australia NOHSC:2011(2003)). This is a comprehensive document referring to how and where to obtain the information as well as approved terms and forms of wording that must be avoided for example the use of N/A, which could mean either not applicable or not available. Such terms should be written out in full. There is globally a consensus system, namely: The Globally Harmonized System of Classification and Labeling of Chemicals (GHS), which is designed to cover all aspects of labeling and classifying hazardous materials including SDSs. An outline of the international standard is included at the end of this chapter.

There are rules governing classification of materials that contain a percentage of hazardous materials. These are given in the criteria for determining hazardous materials and are a function of the degree of hazardousness and the concentration of the ingredient. In the case where there are several different hazardous materials, the various factors are summed up and a total degree established. These factors are also used to determine what risk and safety phases are applicable to the particular material.

The emphasis in writing an SDS should be that it is written in a clear and understandable manner and unnecessary technical terms should be avoided. Use layman's terms rather than complex medical terms.

All SDS documents must be updated every three years and fresh versions issued. With the first delivery or sample of any material, a copy of the SDS must be received and studied before use. The documents must be available to all people coming in contact with the material. The terms used should be fully explained and understood by all.

Governments

Governments at various levels issue rules and regulations controlling many different aspects of the operation of the polyurethane industry. Some of the typical information available includes:

- Uses, sale and storage of MOCA
- Health surveillance of people using MOCA
- Use and storage of solvents
- Use and storage of diisocyanates
- Safety in transport
- Monitoring of diisocyanates
- Format of SDS documents
- Definitions of dangerous and hazardous chemicals
- Workshop safety requirements
- Forklift driving and crane driving
- Others as deemed necessary by local authorities

There are also the local government rules and regulations regarding the location of factories making and processing polyurethanes. These regulations vary from state to state and country to country, so the local requirements must be considered and implemented.

There are standards issued by both the standard organizations (e.g., ANSI, BS or AS) and the professional industry associations regarding the recommended procedures for handling diisocyanates and materials such as MOCA. The Polyurethane Alliance has issued guidelines for the transportation of TDI [1].

Suppliers

Suppliers of polyurethane raw materials issue SDSs for all their products as well as data sheets on the individual materials with additional information about the material concerned. This may be important as in the case of pure MDI where the storage conditions are vital.

Suppliers of bulk raw materials will provide details of handling their material, including details of the transfer procedures from the supply vehicles. The information may also include the suggested correct valves, blanketing and heating if required.

All SDS documents must be read from a practical as well as scientific view point. The most appropriate grade of safety equipment must be used. The equipment must stop the offending material from reaching the person it is protecting. The useful life of the equipment must not be exceeded. Engineering the problem out is the best solution. An example is the fully automated system for handling MOCA.

3.2.2   Signage

Internally all permanent pipe work should be plainly labeled with its contents, for example compressed nitrogen, steam, polyol or diisocyanates. The signs must be clearly visible and in accordance with local regulations and also indicate the direction of flow.

The entrance to the plant must be distinctly placarded with a permanent sign indicating the nature of the dangerous and hazardous materials in the factory with the basic emergency response to each. The "UN number" for each should be indicated on the sign as well as the appropriate emergency and contact number for example the UN number for TDI is UN 2078. Hazard information cards should also be available and sent with each load of isocyanate.

Outside storage containers must also have hazard information relating to their contents attached.

3.3   Prepolymer Preparation

3.3.1   Isocyanate

Isocyanates must be used strictly in accordance with the government's and supplier's recommendations. The material is in the unreacted form with the potential for the presence of vapors. MDI at temperatures of above 50 °C has a substantial vapor pressure compared to when it is at ambient temperature. Care must be taken if work is carried out using PPDI as it has a very high vapor pressure. The fumes will bring tears to the eyes if contact is made.

Isocyanate vapors are present in all isocyanate-containing compounds. The vapors are present to a lesser or greater degree depending on the type of isocyanate and the temperature. The vapor pressure can differ remarkably at room temperature but tends to even out at higher temperatures. Materials such as CHDI have a very high vapor pressure even at ambient temperatures. Isocyanate vapors are a sensitizing chemical that will set off allergic responses both to the skin and to the lungs. They will set off a bronchial/asthmatic attack even at very low levels. Workers should be medically checked prior to work and made aware of the potential dangers.

Warning

If the presence of isocyanates can be smelled, it means that the safe exposure limits have been exceeded.

Engineering controls must be the first line of safety when using any diisocyanate, followed by the strict use of personal protective equipment (PPE).

There also must be strict engineering protocols so that cross-connection between sources of polyols and diisocyanates cannot be made. Personal and atmospheric monitoring must be carried out. It is very important that the operators in this area be under medical screening.

The level of vapors in the air is controlled in many jurisdictions by government regulations. In many situations only the isocyanate (-NCO) level is specified, irrespective of the type. Some typical limits are given below. Check local rules for your limits.

Authority	Exposure	Limits
ACGIH Threshold Limit	Time Weighted Average	0.005 ppm
US OSHA	Ceiling Limit Value	0.02 ppm, 0.2 mg/m^3
Worksafe Australia	TWA	0.02 mg/m^3
Worksafe Australia	Short Term Exposure Limit (STEL)	0.07 mg/m^3

Proper gloves and respirators with fresh isocyanate absorbing canisters should be worn. Engineering solutions are always the best.

Isocyanates are also strong skin sensitizers and can cause irritation to the eyes and mucous membrane. The correct protection equipment must be worn and there must be thorough washing of the hands and arms after exposure to isocyanate liquid or its vapors.

The factory should have the correct hazardous chemical signage at every entrance to the shop floor. Protocols must be set up to handle any spilled isocyanates. Minor spills in-house must be correctly neutralized and the waste disposed of according to local regulations. The local fire and emergency authorities must be made aware of the presence of isocyanates in the factory and have the appropriate documentation of how to handle any emergency. The industry professional body should have this information available for the fire and emergency authorities.

3.3.2   Polyols

Polyols do not have the same extreme toxic properties as isocyanates but must still be treated with care. They are not treated as hazardous but may cause some skin and eye irritation. Basic precautions such as impervious gloves and eye protection must be worn.

Care must be taken when melting any solid polyols not to overheat the drum. This can cause an explosive drum rupture.

Melting Polyols

Steam is often used to melt frozen MDI and PTMEG. The steam generators and associated controls must be maintained according to the local regulations to prevent major problems, including a possible explosion. Steam carries far more energy than hot water. It must be treated with care to prevent severe-burns.

Areas where there are hot drums and pipes must be clearly labeled with appropriate signage. Hot polyurethane prepolymer can cause bad burns, and the potential for free isocyanate vapors is very high.

3.3.3 Prepolymers

Free Isocyanates

Free isocyanates are unreacted diisocyanates left over after the production of prepolymers. The manufacture of prepolymers is normally stopped prior to the consumption of all the diisocyanate. This is so that the material is at the optimum point for further processing. This free isocyanate may either be left in the product (at a level of 0.1 to 0.8% depending on grade) or removed by the manufacturer to a lesser or greater degree. Special low free isocyanate grades (LF grades) are available. These grades have health and safety as well as some technical advantages.

Monitoring must be carried out to check that the levels of isocyanate vapors do not exceed the local limits. These limits vary from area to area. In some areas of the world, only the isocyanate level is specified and is not controlled by the type of isocyanate. Care must be taken as the vapor pressure of different isocyanates may vary by a factor of one hundred or more. Plants that may comply with the atmospheric isocyanate levels when using MDI-based material, may have too high a level when using TDI-based prepolymers.

Certain blocked prepolymers contain chemicals such as methyl ethyl ketoxime. The blocking is on the isocyanate (NCO) groups. When the prepolymer is heated, the vapors of the blocking agent are given off. These vapors can cause irritation if inhaled.

3.3.4 Curatives

There are two main groups of curatives used in the castable polyurethane industry:

1. Amine-based
2. Diol-based

Amines

Amine-based curatives are a popular choice when working with TDI-terminated prepolymers. The two most used amine-based curatives are MOCA and Ethacure 300. Other amine curatives used include Versalink 740M, M–CDEA and Cyanacure. MOCA is technically a very good curative but it is a classified Class 2 suspect carcinogen and is listed in the safety regulations as a chemical along with such materials as asbestos. MOCA must be used under a strict code of practice. The workers using it must be health screened quarterly by a competent medical authority. That means that the level of MOCA in the urine must be tested and the results evaluated by a competent physician in this field.

There is a feeling in the industry that MOCA is not as dangerous as claimed and is just based on old information. There is, however, a large body of information that shows that MOCA reacts with DNA to form an adjunct. There is also evidence that Ethacure 300 has some mutagenic effects but to a much lower degree. The risks involved are of a chronic nature; that means that they are not evident for many years after the initial exposure. It is most important that the code of practice be strictly adhered to. There was a reported case in 1988 of two non-smoking males under the age of 30 who developed bladder cancer. They were exposed to MOCA during its manufacture but had also worked in other sections with another carcinogen.

Ethacure 300 is a very much easier material to handle as it is a liquid at room temperature. It is classed as a hazardous substance according to the safety authorities criteria. As with all chemicals it must be treated with respect and used in accordance with the SDS, which suggests that there should be sufficient ventilation. Neoprene gloves must be worn with disposable or impervious clothing.

Amine curatives that are solid at room temperature need to be melted prior to use. The hot molten material must be handled with care as it can cause severe burns and may be absorbed into the skin.

Data for M-CDEA published by it manufacturer[4] indicates that it is not as hazardous as MOCA or MDA, with lower toxicity and irritation.

Diols

These curatives are normally hydroxyl (OH)-based. Although not as dangerous as the amine-based curatives, these materials are still hazardous, as are all hydroxyl-based materials, with the exception of ethanol in small quantities. Safety precautions such as gloves and impervious clothing should be worn. All chemicals must be treated with respect.

3.3.5   Catalysts

There are three main groups of catalysts used in the casting industry:

1. Organometallic materials (bismuth, mercury or tin soaps)
2. Amine-based material (DABCO 33LV)
3. Organic acids (adipic, oleic acid)

Metallic

These materials provide an extremely fast gel-rate of castable polyurethanes. They are the most hazardous of the three groups. Mercury containing catalysts must not be used in any systems that will come in contact with food. The local limits (if allowed) for mercury-based catalysts must be adhered to.

Metallic catalysts will cause very rapid poisoning if ingested. Some catalysts, for example phenyl mercuric acetate, will result in very strong caustic burning. Contact with the skin will cause a very dangerous burn. Care must be taken with these materials even though they are only used in very small quantities. These catalysts are extremely toxic if digested.

Substitutes

Due to the safety issues in the use of mercury and tin catalysts in both the processing and final product, the use of newer bismuth or zinc catalysts should be seriously considered.

Amine

DABCO 33LV is classified as hazardous according to the criteria of safety authorities. The material has a typical ammonical odor and can be absorbed through the skin. If accidental contact is made, medical attention should be sought after flushing with water. Inhalation of the vapors should be avoided by use of correct engineering practices or a respirator.

Organic Acids

Organic acids such as adipic and oleic are the slowest working of the group but are the least hazardous. Normal procedures should be taken when using them, such as wearing gloves and avoiding the inhalation of any dust. These catalysts are normally used with TDI-based systems.

3.3.6 Other Additives

Plasticizers

Plasticizers such as DIOP, TCP and Benzoflex^{TM}9-88 should be handled with gloves as they may cause minor skin irritations when small concentrations come in contact with the skin. Large-scale exposure to TCP must be avoided as absorption through the skin can cause nerve damage. If this happens, the material must be removed immediately and the skin then washed with copious quantities of water. Medical advice should then be sought.

Nonreactive plasticizers are used to reduce the hardness of the polyurethane. Phthalate plasticizers have a number of health concerns connected with them. Only those allowed in the country where the goods are being made and sold must be used. Phosphate plasticizers/flame retardants must also be vetted prior to use. Some require special government clearance.

Pigments are normally ground in a plasticizer so the same care must be taken. However the grinding medium must not be hygroscopic (like polyols).

Spills of plasticizers must be cleaned up immediately by absorption using standard oil-absorbent granules. These must be disposed of according to local rules and regulations.

The use of simpler phthalates (such as dibutyl phthalate) must be carried out with care due to their higher vapor pressure and greater potential for hazardous health effects.

Silica

Silica is used as a thickening agent in polyurethanes. The grade of silica is a fine anhydrous silica typically sold under the name of AEROSIL®. The particle size is very fine, normally less than 40 millimicrons. The bulk density of the material is extremely low and it is very light and fluffy. Inhalation of the dust must be avoided. From the method of manufacture there should not be any crystalline material in it. Unless properly contained, the very fine material also is a dust problem.

Nanoparticles

By their nature nanoparticles are very small. PPE must be used when handling these materials and if large quantities are used, engineered safehandling procedures should be adhered to.

Pigments

Pigments that are used should not contain cadmium, which has hazardous properties. Pigments will normally be supplied in a paste form let down in either polyol or a plasticizer. Pigments can be very messy and stay on the skin for a period of time if not removed promptly. Disposable impervious gloves should be used when using pigment concentrates.

3.3.7   Solvents

Solvents are found in two main areas, namely directly as a solvent and as a carrier for mold release agents and in bonding paints.

Solvents can damage the skin and respiratory tract. Safety precautions such as wearing an appropriate mask and impervious gloves are required.

The hazards presented by solvents include:

1. Direct health risk
2. Fire hazard

Mold Release

Mold release agents are supplied in either a solvent or in water. The solvent-based materials present handling problems as they are flammable. There are two potential sources of ignition: (1) flames used to pop bubbles that come to the surface of the casting, and (2) arcs from electrical contacts can also set off the solvent vapor.

Water-based mold release agents need more heat to dry the coating and to reheat the reinforcing. Solvent-based mold release agents tend to flash off and leave a coating with greater ease. The operation with the solvent-based material should be carried out away from any flames.

Cleaning Solvents

Only sufficient solvents for immediate use should be kept on the shop floor. The bulk of the stock must be kept in a special flammable storage container.

Besides the fire hazard that they present, any solvent that is spilled on clothing or the skin presents a health problem. Suitable impervious gloves and eye protection must be worn when using any solvent. The contaminated clothing must be removed and the contact area washed well. As the solvents will have removed the natural oils from the skin, they must be replaced by an industrial-grade barrier cream.

Any spills of solvent on the floor should be absorbed into "kitty litter" and disposed of in an approved manner.

3.3.8   Heat

Prepolymer/Curative

When handling heated prepolymers or molten prepolymers, impervious gloves must be worn. These materials can cause bad burns as they are poor conductors of heat. In this condition they can readily be absorbed into the body.

If MOCA is overheated to above 140 °C, dangerous fumes will be given off.

Molds

A hot metal mold will feel very much hotter than a hot polyurethane or epoxy mold. This is due to the thermal conductivity of the mold material. For light metal molds, heat-resistant gloves must be worn. For larger molds, suitable mechanical handling must be employed. The handling equipment must be of suitable size and in good condition.

Dermatitis can result from using a heat glove (nonimpervious) that has been contaminated with either a solvent or polyurethane prepolymer. The same applies to gloves used in the manufacture of prepolymers.

Curing Ovens

Everything inside an operating oven must be considered to be at the set temperature. Large walk-in ovens could have high concentrations of solvent or isocyanate vapors. Proper PPE must be worn when entering them. If too long a time is spent in a curing oven, dehydration can occur.

Many systems require the isocyanate side to be in the range of 70 to 100 °C and over 100 °C with curatives such as MOCA and M-CDEA. A problem arises as a single pair of gloves does not give protection against both heat and the molten curatives. The normal recommendation is to have an impervious glove next to the skin and a heat-resistant glove over this glove. More automated engineering solutions should be looked into as the twin glove is not overly popular.

Products of Combustion

Polyurethane products, including isocyanates, produce dangerous fumes when involved in a fire. Polyurethanes and diisocyanates, like other nitrogen-containing materials such as wool, will produce both carbon monoxide and hydrogen cyanide gas when burned.

Reports [2], [3] on the combustion products of polyurethanes, show that the levels of carbon monoxide in an oxygen-deficient burning condition can reach 290 to 640 mg per gram of sample burned, with cyanide (as HCN) up to 34 mg per gram of sample burned. This data is from laboratory tests and may not represent actual combustion conditions. Professional firefighters should be involved in all situations where there is an uncontrolled fire in a polyurethane plant.

3.4 Engineering

Handling Tools, Materials and Equipment

The most important point is that any tool or jig used must be kept in first-class condition. The main aim with tools and equipment used in the polyurethane shop is that they are there to prevent injury to the workers and to speed up the molding and demolding process.

All safety and protective clothing must be suitable for the application as well as being comfortable to wear. A typical example is that if very cheap safety glasses are provided with non-optical lenses, they will very soon tire the eyes and cause the operator not to use them. The correct balance between safety and operator comfort must always be maintained.

Hazards, wherever practical, should be engineered out.

3.4.1 Gasses

Gasses used in a processing plant will be received either as a cylinder or as a liquid in a cryostat (large vacuum flask).

Cylinders

Gasses that are used in a polyurethane factory include nitrogen, butane or LPG for heating and oxygen and acetylene for welding. Compressed air is used to remove parts from molds and can be generated on site with a small compressor.

All cylinders must be stored in an appropriate manner and properly secured. The supplier of the gas can provide local details of any requirements.

Compressed Gasses

Nitrogen is often used to blanket prepolymers and curatives after the container has been partially used. The blanketing system, if left in place permanently, must be engineered such that the container cannot build up a level of pressure in it that renders the whole situation unsafe.

It must always be taken into account that polyurethane prepolymers are messy materials, and they will cure in the air to form a solid mass. A major precaution would have to be that the gas being delivered to the vessel is at a very low pressure and there is a safe continuous bleeding point.

The pipelines used to convey compressed air or nitrogen must be designed to carry pressure. Low-pressure flexible PVC tubing is not suitable. Where the gasses are carried for long distances, solid metal lines with take-off junctions should be provided. Flexible lines should be kept clear of heated ovens and open flames. Kinks and other mechanical damage should be avoided and when it occurs, it should be immediately repaired.

From a work point of view it is far better to have the compressed air coming down from the ceiling than lying either on the benches or across the floor where it can easily be damaged or trip someone.

Any line, whether it be gas or water that is under pressure and is not secured, will slash around like a snake and can cause considerable damage and harm to a person.

Compressed air should have any free water removed to prevent contamination of prepolymers.

Liquid Nitrogen

In production units where polyurethane prepolymers are used, liquid nitrogen is an option to provide an inert atmosphere in the reactors. The unit should be set up professionally by a supplier and the liquid nitrogen unit properly maintained.

If it stays in contact with the skin, liquid nitrogen can cause freezing burns due to its extremely low boiling point (-195 °C). Evaporating nitrogen can cause the oxygen in the air to freeze out.

It must be remembered that nitrogen when mixed with oxygen (i.e., as in air), is nontoxic but when oxygen is not present, it will cause suffocation.

Any entry into a reactor that has been filled with nitrogen must be fully flushed with air. Only people who are licensed to enter enclosed vessels and are wearing the correct respiratory gear, with qualified backup, may enter the vessel.

3.4.2   Workplace

Electrical

Electrical equipment is the field of qualified electricians; therefore any changes, repairs or modifications should be done by an approved electrician. The overall condition of electrical equipment, including extension leads, must always be observed and any wear and tear replaced or made good by a qualified person. Leads should not be allowed to drape all over the floor or in areas where a hot or heavy mold may be accidentally placed on them.

Lifting Gear

Large molds must only be moved by approved lifting gear. The mold should have eyes attached of size and number to take the weight of the mold. The weight of the mold should be clearly marked on its side. Certified chains and hooks should be used. The capacity of the chain must be sufficient to take the load applied to it.

When lifting the load, it must be kept as close to the floor as practicable to prevent harm in case of gear failure. All equipment must be fully checked and replaced if damaged.

Forklifts

Forklifts must only be operated by qualified personnel and all standard precautions taken. Load limits must always be carefully obeyed, taking into account the effects of high lifts. Speed must be kept to a minimum when working inside the building.

Proper drum-handling equipment, such as drum clamps or parrot beak clamps, must be attached and used when moving drums.

There is poor air circulation in shipping containers; therefore there is a potential for heat and fume build-up, which makes it hazardous when working inside it. All drums must be inspected for leaks or bulging as they are being unloaded.

Machinery

Workplace machinery is used either for the finishing operations on polyurethane castings or in the production of prototypes and molds. All people using the machinery must have had safety training in the operation of the equipment.

All safety guards, where fitted by the manufacturer, must be kept in use. The cutting equipment must be sharp and the clearances suitable for polyurethane. Poor machining practices can cause localized heating and the decomposition fumes are dangerous.

Appropriate PPE must be worn during any operation. These include safety glasses with side shields and a face shield if chips or pieces fly off the machine.

The correct attaching of the piece to the machine is important to prevent accidents. The machine must not be started until you are sure that the work will not become dislodged during machining operations.

The dust from any dry grinding operations must be removed as it is formed. The dust must not be inhaled.

3.5 Safety Data Sheet Format

The data must be presented in the manner specified by The Globally Harmonized System of Classification and Labelling of Chemicals (GHS). The information contained in Sections 12 to 15 are for control by agencies other than Occupational Health and Safety.

The following is the basic layout of the sixteen-point SDS:

1. **Product and Company Identification**

   Product information

   Product identifier

   Manufacturer's name, street address, city, state, postal code
      and emergency telephone number

   Supplier identifier, suppliers street address, city, state, postal
      code and emergency telephone number

   Product use

2. **Hazards Identification**

   Emergency overview

   Regulatory status

   Potential health effects

   Toxicological information

   Route of entry (including skin contact, skin absorption, eye contact,
      inhalation and ingestion)

   Effects of acute exposure to product

   Effects of chronic exposure to product

   Irritancy of the product

Sensitization to the product

Carcinogenicity

Reproductive toxicity

Teratogenicity

Mutagenicity

Names of toxicologically synergistic products

Potential environmental effects

3. **Composition/Information on Ingredients**

Hazardous ingredients

For each ingredient, the concentration expressed as percent weight/weight,
    percent volume/volume or percent weight/volume must be indicated

CAS registry number (for each ingredient) and product criteria
    identification number

Confidential business information (trade secrets)

4. **First-Aid Measures**

First aid procedures

Specific first aid measures

Note to physicians

5. **Firefighting Measures**

Flammable properties

> Conditions of flammability
> Flash point and method of determination
> Upper flammable limit
> Lower flammable limit
> Auto-ignition temperature

Explosion data - sensitivity to mechanical impact

Explosion data - sensitivity to static discharge

Extinguishing media

Suitable extinguishing media

Means of extinction

Unsuitable extinguishing media

Protection of firefighters

Specific hazards arising from the chemical

Hazardous combustion products

Protective equipment and precautions for firefighters

6. **Accidental Release Measures**

   Procedures to be followed in case of a leak or spill

   Personal precautions

   Environmental precautions

   Methods for containment

   Methods for clean-up

   Other information

7. **Handling and storage**

   Handling

   Handling procedures and equipment storage

   Storage requirements

8. **Exposure Controls/Personal Protection**

   Exposure guidelines

   Exposure limits

   Engineering controls

   Specific engineering controls to be used

   Personal protective equipment (PPE)

   Personal protective equipment to be used

   > Eye/face protection
   > Skin protection
   > Respiratory protection

   General hygiene considerations

9. **Physical and Chemical Properties**

   Physical state (i.e., gas, liquid or solid)

   Odor and appearance

   Odor threshold

   Specific gravity

   Vapor pressure

   Vapor density

   Evaporation rate

   Boiling point

   Freezing point

   pH

   Coefficient of water/oil distribution

10. **Stability and Reactivity**

Chemical stability

Conditions to avoid

Conditions under which the product is chemically unstable

Incompatible materials

Name of any substance or class of substance with which
  the product is incompatible

Hazardous decomposition products

Possibility of hazardous reactions

Conditions of reactivity

11. **Toxicological Information**

LC_{50} (species and route)

Effects of acute exposure to product

Effects of chronic exposure to product

Irritancy of product

Sensitization to product

Carcinogenicity

Reproductive toxicity

Teratogenicity

Mutagenicity

Name of toxicologically synergistic products

12. **Ecological Information**

13. **Disposal Considerations**

Waste disposal

14. **Transport Information**

Basic shipping information

Special shipping information

Additional information

15. **Regulatory Information**

Safety data sheets should include a statement such as:

"This product has been classified in accordance with the hazard
criteria of the country of issue and the SDS contains all of the
information required by those regulations."

It is advisable to include the WHMIS classification here as well.

16. **Other Information**

> Preparation information
>
> Name and phone number of the group, department or party
> responsible for the preparation of the SDS
>
> Date of preparation of the SDS

References

[1] API. TDI Transportation guidelines. *Trade literature brochure*, pages 1–35 plus Appendices, 2002.

[2] E. A. Boettner, G. L. Ball, and B. Weiss. Combustion products from the incineration of plastics. Technical report, University of Michigan for U. S. Enviromental Protection Agency, Ann Arbor Michigan, 1973.

[3] DuPont. Combustion products of a urethane vulcanizate cured with MBCA. page 1, 1981.

[4] Lonza. Lonzacure M-CDEA — The superior curative for the polymer industry. *Trade literature brochure*, pages 1–7, 1992.

Part III

Processing

4

Prepolymer Production

4.1 Prepolymers

Prepolymers are formed by the reaction of an excess of diisocyanate with a polyol. One of the isocyanate groups (NCO) reacts with a hydroxyl group (OH) of the polyol. Another isocyanate group reacts with the second OH group. A very important feature of this reaction is that there are no by-products formed. The resultant product has isocyanate groups on both ends with urethane bonds between the original polyurethane and the diisocyanate group.

2 Diisocyanates   1 Polyol

Prepolymer

where X is either an ether or ester group.

Prepolymers are normally produced with a mole ratio of approximately two moles diisocyanate to one mole of polyol. This is the representation of a straightforward commercial prepolymer. The ratios are often varied to give enhanced different properties. The composition of the polyols is also varied for similar reasons. An example of this is the addition of a triol in hard compounds to improve the compression set. Blends of PPG and PTMEG are also often used to improve the properties of the classical PPG-based prepolymers.

Basic Molecular Weight and %NCO Values

The following are theoretical NCO values for a range of different molecular weights.

Polyol MW	TDI	MDI
400	11.2	9.3
650	8.4	8.4
750	7.7	6.7
1000	6.2	5.6
2000	5.6	3.4

In commercial practice the ratios may be altered slightly to enhance certain properties. The NCO levels will also change depending on the purity of the isocyanate as well as the processing conditions.

There are several reasons to work through the prepolymer route as opposed to the one-shot technique. These include:

- Ease of production
- Lower isocyanate vapor levels
- Designed structure
- Exotherm of final reaction reduced
- Easier handling of components
- Final properties of system

If the ratio is very much larger (e.g., four to one), the resultant product is called a quasiprepolymer.

4.1.1 Ease of Production

The preparation of prepolymers and quasiprepolymers allows for the production of polyurethane parts by component manufacturers without the large capital outlay required to produce materials from the basic raw materials. The production of any prepolymer requires a good understanding of the chemistry involved. The final quality of the polyurethane product is dependent on the initial control of the chemistry of the system and would be expensive for small operators to carry out.

4.1.2 Isocyanate Levels

Free isocyanate vapors in the workplace are strictly controlled because they are a health hazard. In the commercial production of prepolymers the diisocyanates can be transferred in a closed system and the final prepolymer stripped of most of the free diisocyanate. Although the final prepolymer is strictly a diisocyanate, its molecular weight is so high that its vapor pressure is very low.

4.1.3 Structure of Polyurethane

Using controlled reaction conditions such as the temperature profile and rate and time of the addition of polyols, more uniform materials can be produced. The correct spacing of the hard segments that is required to produce the physical properties, can be obtained.

The controlled conditions will also help prevent the formation of undesirable side products such as allophanate, biuret and trimers. These reactions will give branching of the polymer chains.

The reaction rates of the various ingredients must be understood to follow the probable course of the reaction. Certain polyols may have secondary hydroxyl groups that are very much slower to react than the primary hydroxyl groups. The preparation of a prepolymer will allow the secondary hydroxyls to fully react. The reaction of a 100 percent 2,4-TDI is much faster than the standard 80:20 TDI isomer blend.

The reaction rate between the terminal NCO and chain extender can be adjusted to allow the prepolymer mix to fill the mold and entrapped air to escape before the material gels.

4.1.4 Reduction in Final Exotherm

The production of a polyurethane elastomer gives off heat when urethane and urea bonds are formed. This will cause shrinkage in the final part. The production of a prepolymer will allow part of the exotherm to be dissipated prior to the formation of the final elastomer. This will in turn result in a lower final exotherm and decreased shrinkage.

4.1.5 Easier Handling of Components

In a number of applications, pure MDI must be used. This chemical is solid at ambient temperature and melts at 38 to 40 °C. The molten MDI is very prone to dimerization. These two factors make it very difficult to use. The normal storage of this material is to keep it below −18 °C.

Dow Chemical [3] gives the following data regarding Isonate^{TM}125M, which is a pure MDI. Table 4.1 details the storage life of pure MDI.

A prepolymer prepared with pure MDI does not have the dimerization problems of the diisocyanate.

In the preparation of elastomers, well-defined hard segments where hydrogen bonding can develop are required. To produce the segments, low molecular weight diols and diamines are often used. These are not overly soluble in the high molecular weight polyols. They can be reacted with some diisocyanate to produce a prepolymer. This prepolymer is more compatible,and therefore less vigorous agitation is required to ensure the proper dispersion of the diols.

TABLE 4.1
Storage Life of Pure MDI

Temperature(°C)	Storage Life (Days)
−17.8	300
−12.2	210
−4.4	68
10	33
25	See note
40.6	31
43.3	35
46.1	35
48.9	28

Note: "Temperature ranges between 21.1 and 37.8 °C give a minimum dimer growth of 0.0114 percent per day. There are wildly erratic results in many recorded instances".

4.1.6 Final Properties

As the molecular weight of the polyurethane chains increases, the physical properties will change. Preparation of a prepolymer will allow the formation of an initial chain that can be extended to its final length with a desired chain extender (curative). The heat given off during curing will not be as great as when the whole reaction is carried out in one step. This will reduce potential problems in the final product.

4.2 Laboratory Preparation

Important Warning

Before commencing any experimental work, the relevant up-to-date SDSs must be studied and the appropriate safety requirements observed. There are often local or in-house regulations for handling certain materials such as MOCA. These must be strictly observed. All people in close proximity to any isocyanate must take special care not to inhale any traces of its vapor as these may cause respiratory problems.

If available, the analysis of the raw materials should be studied especially with regard to purity, potential interfering contaminates, functionality, moisture and acidity/alkalinity. The molecular weight is often not given directly but must be calculated from the OH value and the functionality of the polyol. (See Appendix D for formula.)

4.2.1 Equipment

Drying System

Polyols such as PTMEG may be a solid at ambient temperature and must be heated to allow further processing. This may be carried out in a convection oven set to 60 °C. It must be remembered that there will be a pressure buildup in the container. The bung may be cracked slightly to vent the gasses given off.

The polyol must be dry prior to use. This is best carried out by heating the material under vacuum at 90 °C for one to two hours to remove any moisture. It must be allowed to cool in a dry atmosphere, for example under a blanket of nitrogen gas.

Figure 4.1 illustrates one form of a typical reaction vessel. The size and type will depend on the availability of equipment.

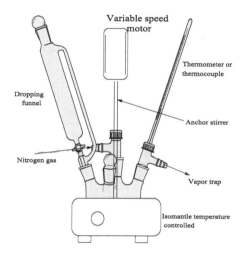

FIGURE 4.1
Laboratory-scale reactor.

For laboratory-size experimentation, the most suited vessel is a wide-necked flask with a clamp-on lid. This style has several advantages in that fitting an anchor stirrer is much easier. The discharging of the final product is relatively quick with this type of vessel. When a batch has been prepared, all parts that have been in contact with the polyurethane must be thoroughly cleaned and dried. Polyurethanes are a very effective adhesive and can be formed by the reaction between the resin produced and the moisture in the air.

Heating

The heating of the reaction vessel may take one of several forms:

- Heating mantle
- Oil bath
- Infrared heaters
- Sand bath

These heaters may be controlled by a thermocouple placed in the liquid coupled to a commercial temperature controller, preferably one that has ramping facilities. The application of heat must be able to be stopped and some form of forced cooling applied if the exotherm becomes too large.

4.2.2   Reactions

The aim when making prepolymers is to have all possible terminal groups be an isocyanate (NCO). By adding the diol to the isocyanate, the reaction will be directed toward terminal isocyanate groups. This is needed when the chain is extended in the final preparation of the polyurethane.

Most polyurethane systems tend to cling to the glass surface so an anchor-style stirrer is the best to use. This provides a sweeping motion that helps prevent a very viscous layer from forming on the surface of the vessel.

The desired weight of diisocyanate is added to the flask plus any ingredients to either adjust the acidity or to catalyze the reactions. See Appendix D for the required calculations. The diisocyanate is heated to the start temperature with gentle agitation. It is best to provide a complete blanket of nitrogen in the reaction setup from this point on.

The polyols are added at a rate so that the exotherm remains under control and the maximum temperature is such that side reactions are reduced to a minimum. Typically the temperatures should be within the range of 60 to 90 °C. As polyols are often viscous, care must be taken to ensure that the correct amount is added to the diisocyanate. This either means a long drainage time or by weighing the amount added to the reaction.

Once all the ingredients are added, the polyurethane must be held at temperature while the reactions proceed to completion. This may be approximately one to two hours.

When more than one polyol is used, the structure of the final product can be controlled by the order in which the polyols are added. The varying reactivities of the isocyanate groups will also affect the structure of the final product.

On completion of the reaction, the material can be carefully degassed to remove any entrapped gasses and unreacted diisocyanate. Care must be taken to prevent foaming. Applying vacuum to the glassware does have the potential for it to implode. Suitable safety precautions must be taken.

Table 4.2 indicates some typical time/temperature reaction conditions for some simple prepolymers.

TABLE 4.2

Typical Reaction Conditions

	System One	System Two	System Three
Isocyanate	MDI	TDI	MDI
Initial temperature	55 °C	40 °C	55 °C
Diol	PTMEG	PTMEG	PPG
Initial temperature	55–70 °C	40 °C	45 °C
Final temperature	80–84 °C	80 °C	80 °C
Hold time	1–2 hours	1–2 hours	1–2 hours

Care must be taken when using polyols that have been made using a strong alkali as a catalyst. These are the standard polypropylene glycols (PPG polyols) made with a potassium hydroxide catalyst. Even though the level is low, the alkali is very reactive. There will be a point in the reaction when all the acidity of the diisocyanate is used up and the product is slightly alkaline. The excess alkali (base) will catalyze the reaction. This will not only speed up the reaction but also cause side reactions and the potential for gelation. The acid/alkali balance of the ingredients should be examined and a final acid excess of approximately 0.33 microequivalents per gram obtained. The acidity may be obtained by adding an organic acid such as benzoyl chloride or 85% phosphoric acid. (See Appendix D for typical calculations.)

4.2.3   Monitoring Reaction

The progress of the reaction can be followed in several different ways.

The classical titration of the total available NCO level can be carried out. Initially all the NCO will be from the diisocyanate. When chain extension occurs, the NCO will be converted to urethane groups and the available NCO will decrease. The titration will have to be carried out rapidly to obtain a reasonably true result. The dry solvent used may have to be adjusted to give quick solubility. A general method is given in Appendix E.

The reaction may also be followed by using a micro-temperature-controlled cone and plate viscometer. As the polyurethane chains increase in length, the viscosity will slowly increase. If side reactions take place, the viscosity will start to increase rapidly. This increase in viscosity is illustrated in figure 4.2.

Using Fourier Transform Infrared (FTIR) spectroscopy, the reaction can be monitored. The progress of the reaction can also be followed by studying the reduction of the primary hydroxyl peaks in the mid-infrared spectra. The urethane band with the approximate wavenumber 1739cm^{-1} will be forming. The NCO band at 2273 cm^{-1} will decrease from a maximum and then stabilize as the chain extension proceeds.

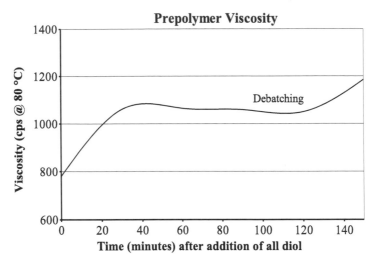

FIGURE 4.2
Final viscosity buildup.

4.2.4 Prepolymer Storage

On completion of the reaction the material (if not immediately used) must be stored in a clean dry container. The material of construction must be able to withstand the heat of the material plus any further reheating required. If metal containers are used, they must be lined with an epoxy or similar lining that is nonreactive to the polyurethane prepolymer. Any remaining air in the container must be replaced with dry nitrogen gas. All mating surfaces of the closure must be free of the polyurethane prepolymer.

4.3 Commercial Preparation

In scaling up from laboratory-size to large-scale production, the same chemistry takes place but the equipment used becomes far more complex. The capital costs are much larger and the financial consequences of a failed batch will run into thousands of dollars.

There are several major components in the production unit:

1. Safety considerations
2. Raw material storage
3. Raw material preparation
4. Reactor

5. Nitrogen

6. Heating/cooling units

7. Process control

8. Vacuum

9. Discharging batch

10. Quality control and assurance

4.3.1 Safety

Safety requirements vary from state to state and from country to country. All local and national requirements need to be checked and taken into account prior to and during any large-scale production. The current SDSs for all chemicals must be available and all warnings observed.

TABLE 4.3

Isocyanate Vapor Pressure Variations

Isocyanate	Vapor Pressure mbar	Temperature °C
Toluene diisocyanate (TDI)	3×10^{-2}	20
Methylene diisocyanate (MDI)	$<1 \times 10^{-5}$	20

Isocyanates have acute and chronic safety problems. Elimination of isocyanate vapors must in the first instance be carried out by engineering methods. The ventilation of the area must be good. Transfer of the isocyanates must, wherever possible, be automated so that there is no human contact. If there is handling by humans, the proper protective equipment should be worn. The personal protective equipment must include safety goggles, impervious gloves, vapor masks, and disposable overalls. If the levels of vapors are expected to be high, a self-contained breathing hood must be used.

Splashes of isocyanate on the skin must be removed by copious flushing with water, then washing with soap, followed by an application of barrier cream. Any splashes to the eye must be washed out very well and medical advice obtained.

The level of isocyanate vapors in the air is specified by local authorities and refers to all isocyanates and not any individual isocyanate. Table 4.3 shows how the vapor pressure of the isocyanate varies from type to type as well as at different temperatures.

Authorities in different countries have different exposure limits. Some of these are illustrated in Table 4.4. It is advisable to find the local limits applicable for exposure to isocyanate vapors.

Storage and production areas that contain isocyanate should have containment walls (bund) that will hold any spilled isocyanate. Spills of solid or liquid isocyanates must be cleaned up immediately. Oil-absorbent granules

TABLE 4.4
Occupational Exposure Limits

	Limit	Reference
TLV	0.005 ppm as TWA :(skin)	ACGIH 2004
MAK	0.05 mg/m^3	
PEL	0.2 mg/m^3 (0.02 ppm)	OSHA
TWA	0.05 mg/m^3 (0.005 ppm)	NIOSH
	0.2 mg/m^3(0.020 ppm)	REL: 10-minute
Sensitization of respiratory tract and skin		(Sah)
IDLH	75 mg/m^3	NIOSH

(kitty litter) can be used to contain the spill and one of the following or similar decontaminants used to neutralize the isocyanate. The material should be placed in a drum and allowed to react for a full forty-eight hours.

If the spill is large, the staff must be evacuated and the emergency services called. All these procedures must be fully documented and the staff trained in their implementation.

Liquid Decontaminant (Nonflammable)	
Water by volume	90%
Nonionic detergent (100%) by volume	2%
Concentrated ammonia by volume (specific gravity 0.880)	8%

Solid Decontaminant	
Sawdust by weight	20%
Kieselguhr, technical or China clay, or Fuller's earth by weight	40%
Liquid decontaminant by weight	40%

Workers may become sensitized to isocyanates, therefore they should be clinically monitored. An initial respiratory check should be carried out prior to commencement of work in the area and then at regular intervals after that.

The air around the production area must be vented out, flowing in the direction away from the workers. The isocyanate level should be monitored by an isocyanate detector such as the Remote Intelligent Sensor (RIS) by AFC International, Inc. Detector badges can also be used.

In designing a reactor, the vessel must be able to withstand both vacuum and a very slight positive pressure. The reactor must be fitted with a blow-off valve to vent any excess nitrogen padding gas. In case the reaction runs out of control, the lid of the reactor must also be fitted with a pressure rupture disk. The fumes and product must be trapped safely.

Entry into the reactor must be strictly controlled and only carried out by suitably qualified and certified workers after complete flushing of the vessel with air. Air breathing equipment and safety equipment must be used. The power supply and feed valves must be safely isolated.

Auxiliary heating equipment such as steam generators must be maintained to the local safety requirements.

When preparing a batch of material, the reactor and the prepolymer are hot. Suitable heat-resistant gloves must be worn. It may be necessary to wear a pair of impervious gloves under the heat-resistant gloves to give both chemical *and* heat protection.

4.3.2 Raw Material Storage

The bulk raw materials such as polyols, isocyanates and cleaning solvents are normally received in one of three different ways:

1. 200-liter drums
2. 1000-liter IBC containers (Intermediate Bulk Containers)
3. Bulk delivery

The main aim of storage is to keep the material in the same condition in which it was dispatched from the raw material manufacturer. All the raw materials must be kept dry and away from extremes of heat. The temperature at which isocyanates are stored is of vital importance. Pure MDI needs to be kept frozen (-14 °C) until just before use. This is to prevent the formation of dimers. The suppliers' recommendations must be taken into account.

If 80:20 TDI is stored below 15 °C, the material will partially crystallize. It can be melted again but because the 2:4 isomer solidifies first, the material must be well mixed prior to use.

Chemical Incompatibilities

Small items such as organic acids (e.g., benzoyl chloride) and catalysts must be kept secure. They can be harmful if they are spilled. Their hazards include severe skin burns and damage to the eyes and respiratory systems. The SDSs for all items must be studied.

Chemicals such as ammonia, amines and water must be kept away from all types of isocyanates as they can react violently.

When receiving bulk raw materials, the grade of the material must be carefully checked before it is transferred to any tank. Besides visual and paper checks it must be impossible from an engineering aspect to add an isocyanate to a polyol tank or polyol to an isocyanate tank as there may be an extremely severe reaction.

General

On receipt of the raw material the certificate of analysis must be checked. Any variations in molecular weight or NCO level must be evaluated and taken into account when the material is used.

It is good practice not to open any drums prior to use. Any partial drums should be flushed with nitrogen gas prior to resealing. All storage must work on a first-in, first-out basis.

The storage area must be secure and have large enough bunded (diked) areas to hold the material if a leak develops. Heating must be provided to the tanks if the local conditions are such that the material may become too cold. This could result in the material being too viscous to pump or for crystallization to take place.

4.3.3   Raw Material Preparation

Prior to the commencement of any production, all state and local regulations must be adhered to as the operations may be hazardous from both a mechanical and chemical aspect. Safety and emergency protocols must be drawn up and checked out. All safety equipment must be thoroughly checked before commencement of any work.

The raw materials used must be kept in clean dry surroundings at an appropriate temperature. The fitting used for isocyanates and polyols must be noninterchangeable. This is to prevent any unexpected reactions taking place with drastic consequences.

The quantity of raw materials kept in stock must be both a function of the straight economic cost and the storage life of the materials. The disposal cost of emptied containers must be taken into account. Many containers cost money to dispose of in a safe and legal manner, especially those that have contained isocyanates.

For all raw materials, the current Safety Data Sheets and any other handling and storage information must be obtained from the raw material supplier, studied and kept on hand. The recommendations contained in this data must be complied with.

Raw materials received in damaged containers must be isolated and rendered safe as soon as possible. Drums with a puncture hole in the side (for example, from a forklift prong) must be laid on its side with the hole on top. The hole can then be sealed and any spilled material neutralized. The pressure in a drum with a bulging end must be carefully released, taking all precautions. The material should be returned to the supplier for correct disposal.

Bulk supplies of liquid isocyanates and polyols must be delivered into their own bulk storage containers. The size of the tank must be large enough to take the new consignment while allowing for some of the previous batch to be present. There must be sufficient ullage to cover temperature expansion. Guidelines for the receipt of bulk TDI are given in supplier and organizational guidelines[1].

The tanks must have a regulated nitrogen gas pad to prevent any moisture from entering and reacting with the material. A pressure relief valve must be built into the system to enable excess pressure to be relieved when the tank is being refilled or the material becomes hot.

If the ambient temperature falls too low, gentle heating must be provided to the tank to keep the material liquid. TDI starts to crystallize at 15 °C [5]. The lines to the processing reactors may also have to be heated. A low temperature may either cause the material to become too viscous or to freeze. The transfer pump must be set up to be able to recirculate the material to help keep the temperature constant.

The isocyanate and polyol storage units must not share pumps or any fittings. The fitting must be such that there is no way that the wrong material can be added to a tank. All tanks must be clearly labeled with their contents and the appropriate safety signs, for example, isocyanate hazard signs. Most regulatory authorities will require that the tanks be in a suitably sized bunded (diked) area.

The transportation and storage of pure MDI must be carefully controlled. For transportation between a depot and the factory, the MDI may be kept at −18 to −4 °C for up to two days. For long-term storage of approximate 300 days, the MDI must be held at −18 °C. This is the total of the warehouse and factory storage times. If MDI must be held in the liquid form it must be kept at 43 to 48 °C for a maximum of four weeks. These requirements are to reduce the formation of dimers.

The most approved method for melting solid MDI prior to use is a drum roller in steam[4] or hot air[2] at atmospheric pressure. Key points:

- Drums free from dents
- Previously opened drums flushed with nitrogen gas (dew point −40 °C)
- All bungs tightly closed
- Rolling speed 5 rpm
- Atmospheric steam pressure or hot air
- Normal time to reach 70 °C quickly is four to five hours

The aim is to melt the material quickly but evenly with minimal formation of dimers.

Polyol can either be a liquid or a semisolid at ambient temperature. These materials will have to be kept in heated bulk containers. Drum stock will have to be melted in a similar manner to pure MDI. To ensure that the material does not absorb moisture from the air, the drums must be blanketed with nitrogen and the bungs properly closed. Table 4.5 shows the melting range of some common diols.

The higher the molecular weight, the higher the melting point and the more viscous the material.

The polyol needs to be heated to provide a good flow of material and to have no tendency to solidify in the feed pipes. A maximum temperature of 70 °C is normally recommended to prevent too high an exotherm.

Stabilizers are added to the system to prevent side reactions and gelation. The aim is to have the final system very slightly acidic (0.33 microequivalents per gram). The two most popular materials are benzoyl chloride and 85%

TABLE 4.5
Melting Range of some Common Diols

Diol Type	Molecular Weight	Melting Point °C
Esters		
Polyethylene adipate glycol	1000	45
	2000	50
Ethers		
PPG	4000	Liquid at ambient
PTMEG	1000	14–23
	2000	26–30
Caprolactones		
Polycaprolactone glycol	830	35–45
	2000	45–55

phosphoric acid (see Appendix D for calculations). The degree of alkalinity of the polyol will depend on the chemistry of the polymerization. If an alkaline catalyst such as potassium hydroxide was used to make the polypropylene glycols, the alkalinity may be relatively high. Some new polyols such as the Bayer Acclaim® range has some phosphoric acid added.

For the best effect, the stabilizers must be added to the reactants before the point of acid/base equilibrium is reached.

Catalysts are normally not added to the prepolymer as they may be rather basic and may cause the reactions to proceed too swiftly.

4.3.4   Reactor

Design

A typical reactor is illustrated in Figure 4.3.

The reactor needs to be built from a grade of stainless steel that can withstand mild acidic conditions. Traces of hydrochloric acid may be present in the isocyanates and in some cleaning solvents. The side and bottom of the vessel must be fitted with either heating coils or a suitable jacket. Depending on design considerations the jacket may either be plain or dimpled. If at all practicable the lid should be completely removable. This is important for major cleaning or if a batch should gel. It does, however, make the maintaining of positive pressure or vacuum more difficult.

The lid of the reactor will have ports to cover the following functions:

- Isocyanate in
- Polyol in
- Outlet port
- General-purpose port (for minor ingredients)
- Nitrogen in

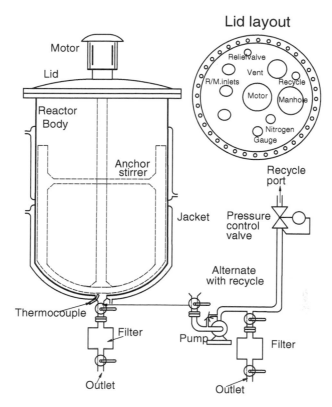

FIGURE 4.3
Factory-scale reactor.

- Vacuum
- Pressure rupture disk
- Low-pressure relief valve
- Pressure gauge
- Recycle port (if needed)
- General viewing (entry manhole)
- Agitator gland

Thermocouple probe pockets need also to be fitted. Their location needs to be in conjunction with the design of the agitator and the need to obtain, as accurately as possible, the true temperature of the prepolymer as it is being made. The lowest probe must be low enough to measure the temperature of the initial isocyanate that is added to the reactor.

The main outlet of the vessel should be at least 75 mm in diameter to allow for a fast flow of the prepolymer. The outlet can either be connected to a recirculating pump or directly to a filter prior to the drum filling. The pump must be capable of pumping the most viscous material quickly. Another key point to be noted when selecting a pump is that it has to have suitable glands. It must also be able to withstand the cleaning solvents and be readily stripped for clearing any buildup in the pump.

Due to the physical size of the reactor, it is normally mounted with the top of the reactor at a mezzanine floor level. This allows for easy access to the inlet ports. The reactor can be mounted at this point on a number of load cells. The load cells must be of such capacity to take the whole weight of the reactor when it is full of prepolymer. In this case all connectors to the reactor must be flexible.

Agitation

The agitation of the material is important, as quick and efficient mixing is required to allow the reaction to proceed correctly. The speed must be such that a vortex is not formed and gasses entrapped in the mix because then they would need to be removed at a later stage. Anchor-style mixers have proven popular in this application.

Keeping the walls clean of buildup is important as this will cause heat transfer problems as the production proceeds. The polyurethane prepolymer builds up in viscosity as the chain length increases and tends to hang onto the wall of the reactor. The heat of the wall will increase the speed of the reaction and hence increase the reaction rate and viscosity. Viscosities in the reactor would be generally between 30 and 4000 MPa.s at a temperature of approximately 90 °C.

The surface of the agitator must be as smooth as possible to prevent unnecessary hang-ups but must also be strong enough to take the torque required. The design and sizing of the motor is best carried out by a company that is experienced in these matters.

As the unit will work under vacuum and a slight pressure, the sealing gland must be able to take both vacuum and pressure. As polyurethane is a very effective adhesive, the seals must be readily replaceable. If a mechanical seal is used, great care must be taken to prevent any product from coming onto a contact surface.

Dispensing Ingredients

The addition of the correct quantity of each ingredient is important. The dispensing equipment must be suitably sized to cope with the various weight ranges involved.

The entire reactor may be mounted on load cells and the weight indicated on a visual display with outputs to logging and other control units. The weight of the reactor with agitator and motor may be in the order of tons. The weight of the heating and cooling jackets must also be taken into account. If a combined heating and cooling system is used, there will be a difference if the jacket has steam, condensate or water in it.

The initial isocyanate may be weighed directly into the reactor and any acidity controlling agents separately added by volume or by weight using a more sensitive weighing system.

The system for dispensing the polyols must be flexible to enable both the rate and the quantity to be varied. The addition of the polyol may have to be slowed down or stopped and restarted if the reaction becomes too vigorous and the temperature increases. As the addition will be over one to two hours, it is also important to keep the temperature of the polyol constant.

4.3.5   Nitrogen

Nitrogen is used to blanket the reaction and the raw materials to stop the absorption of moisture and to prevent oxidation of the product.

The dew point of a gas is the temperature at which the water vapor present in the gas saturates the gas and begins to condense. (Dew points below 0 °C are sometimes called the hoar-frost points). Industrial-grade nitrogen gas normally is dry enough for use. (Dew point −40 to −50 °C)

In these applications nitrogen is needed in volume but at a very low pressure, slightly above atmospheric. The supply must be set up to prevent over-pressurization of the containers. Suitable pressure relief valves must be installed to vent any excess pressure.

There are three main routes to supply nitrogen to the system:

1. Compressed gas cylinders
2. Evaporated liquid nitrogen
3. On-site preparation

Compressed gas cylinders have the lowest capital cost of the three systems. Cages of four or more cylinders are supplied as a group and the system connected to the cylinder manifold. The gas cylinders are supplied at a high pressure and the gas must be reduced to just above atmospheric pressure for use.

Liquid nitrogen is stored in large vacuum vessels (cryostats). The liquid nitrogen is evaporated through a heat exchanger and the gas used in the factory. The units are normally supplied and maintained by the gas company. The cryostats can be easily topped up by a road tanker. This system is more economical in medium to large units.

Dried compressed air can be passed through molecular sieve tubes where the nitrogen is separated from the oxygen and inert gasses present in the air. The advantage of this system is that the production of the nitrogen is on site. The rate of take-off will affect the final purity of the gas. The units will have to be sized for the application. A buffer may have to be built into the system. This means that output from the unit is compressed into gas cylinders to buffer the peak demand.

Nitrogen is an "on cost" but is of vital importance for a quality product. Removal of air from the reactor by vacuum will save on the use of nitrogen to replace all the air.

4.3.6   Heating/Cooling

The temperature of the reactor needs to be controlled for several vital steps in the reaction:

- The reactor needs to be brought to the start temperature.
- The isocyanate needs to be heated to the initial temperature.
- Heat must be available to raise the temperature to the final holding temperature.
- Cooling must be available to cool the reactor if the temperature rises too high.

Heating may also be required to keep the polyols fluid and to melt pure MDI if used.

There are two major alternatives in the heating, namely steam or a heat transfer medium such as Dowtherm[TM][1] heat transfer fluid. The system must be such that the heating medium will be rapidly replaced by a cooling medium. The switching must be on a fail-safe sequence.

If steam is used for heating, all pressure must first be vented safely prior to water being used to cool the reactor. Steam is introduced at the top of a jacket and the condensate released from the bottom. Water normally flows from the bottom to the top. The moist air needs to be vented prior to circulation.

If a pump is fitted to the discharge of the reactor, it can be used to fast circulate the hot prepolymer through cooling tubes to rapidly reduce the temperature.

When the temperature probes indicate a too rapid temperature rise, the addition of polyol must be slowed right down. If the temperature still rises too fast, the addition of the polyol must be stopped until the temperature is brought back under control.

The temperature probes must be positioned so that the temperature of the prepolymer indicated is unaffected by the wall temperature of the reactor. A second thermocouple positioned elsewhere in the reactor can be used to detect runaway reactions.

Dow Chemical [4], in their pamphlet on prepolymer production, gives a formula to calculate the total heat given off by the instantaneous reaction of polyol with an isocyanate. Figure 4.4 indicates the instantaneous heat rise generated when polyols of different molecular weights are reacted with TDI and MDI.

The slow addition of the polyol will allow heat to be dissipated and thus the full rise in temperature will not be realized.

---

[1] Dow Chemical Company.

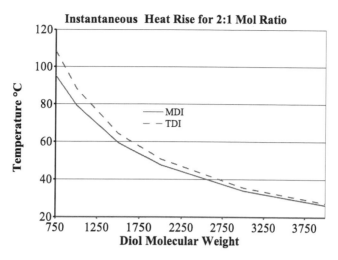

FIGURE 4.4
Instantaneous heat rise for a 2:1 mol mixing of an diisocyanate and diol.

4.3.7 Process Control

The aim of the factory production is to produce, as economically as possible, a consistent product with the specified properties. The key points in achieving this are:

- The correct weight of each ingredient added
- The addition in the correct order
- The correct rate of addition
- The heat history of each batch is the same
- Minor variations in raw material values are accounted for

A set procedure for each grade must be set up and strictly adhered to. For normal quality assurance, the weights, temperatures and times must be logged. The degree of sophistication of each plant can vary from a purely manual system to a computer-controlled, fully automated unit.

Temperature control is normally carried out using thermocouples in a stainless steel pocket. The type of thermocouple used is either a platinum resistance detector (RTD) or a thermocouple using two dissimilar metals that produces a voltage (EMF). The indicators for these thermocouples must match the probe type and grade. The positioning of the probes is very important as is any lag (delay) in the system. The output from the probe is connected to the indicator and/or controllers. Most indicators have at least a set point with an on/off output. The more advanced units will allow anticipated switching, more than one set point, temperature ramping between temperatures, and time and

hold facilities. Thermocouple break and over-temperature alarm outputs are
also commonly provided features.

The output from the temperature controller can be fed into control valves
either on/off or proportional or into a PLC that can be programmed to turn
off the polyol feed and heat and start the cooling cycle.

The amount of material added can either be measured by weight or by
volume. If the material added is by weight, the size and capacity of the scale
must be taken into account. Electronic scales are based on the number of
counts in a given range. Typical figures may be 1 in 3000, 1 in 10000 and 1 in
20000. For a given range, these will control the readability of the scale. The
readability is illustrated in Table 4.6.

TABLE 4.6
Readability of Scale Ranges

Count	1 in 3000	1 in 10000	1 in 20000
Load cell range			
0 – 300 kg	0.1 kg	0.03 kg	0.015 kg
0 – 1000 kg	0.3 kg	0.1 kg	0.05 kg
0 – 6000 kg	2.0 kg	0.6 kg	0.30 kg

Scales currently in use are usually based on load cells. Modern weight in-
dicators/controllers have outputs that can be connected to valves that control
the flow of the material. Batch weights can also be stored in some units.

When the material is measured out by volume, the density of the feed
must be known as the weight is a direct function of the density and volume:

$$Density = \frac{Weight}{Volume}$$

$$Volume = \frac{Weight}{Density}$$

The density at any given temperature will have to be obtained either
experimentally or from the raw materials supplier.

4.3.8 Vacuum

Reducing the pressure of the reactor serves three main purposes:

1. Removal of moist air/gasses from the reactor
2. Removal of isocyanate vapors after the batch is complete
3. Removal of any gasses in the prepolymer

The reduction of pressure in the reactor will decrease the boiling point of
any unreacted chemicals in the batch. This is increased with elevated temperatures. Care must be taken to prevent foaming. Apply vacuum slowly and
release if required to break the foam.

Direct contact of the isocyanate vapors to the pump must be avoided. The fumes must either be frozen out with a liquid nitrogen trap and/or a trap containing a plasticizer such as Messamoll®.

The resistance of the materials of construction and the seals of the pump must also be resistant to isocyanate vapors. The final exhaust fumes must be vented well away from all operations. Valving must be installed to prevent pressure in the system from being forced into the vacuum system.

4.3.9 Discharging Batch

When the batch is complete, the material must be discharged as soon as practical into the dispatch container. The viscosity at this point will be low to assist in flow.

The drums must be clean and dry before being filled. The inside of the drums must be coated with either a phenolic or epoxy resin to provide resistance against attack from the prepolymer. The wall thickness must be sufficient to prevent collapse due to a vacuum being formed as the material cools and contracts. Prior to sealing the drum, nitrogen should be introduced into the free space in the drum.

IBCs (intermediate bulk containers) must be clean and dry and made from material that can resist the temperatures when filling the container. The discharge temperature may have to be lowered to cope with the initial temperature of the prepolymer.

If a recirculation pump is fitted to the reactor, the material may be discharged from the pump circuit. The valving must be such that the material is recirculated when not being discharged. Pressure relief valves must also be fitted to the pump.

The prepolymer may also be discharged by gravity into the container. Nitrogen gas pressure may be used to assist the flow of prepolymer.

The prepolymer is dispensed by weight into the containers. The scales used must be trade certified and filled to the specified weight. This may be carried out either manually or automatically by controlling output from the scales.

Depending on the smoothness of the reactor and heat uniformity, solid polyurethane is sometimes formed and may be discharged with the material. These lumps must be removed by passing the material through a free-flowing filter. The filter material must be resistant to both the temperature and prepolymer. Polypropylene filter material has been successfully used. The filters must be cleaned after use.

The reactor needs to be kept clean to keep the heat transfer optimal and to prevent solid material in the prepolymer. The method employed is to use an appropriate solvent such as methyl ethyl ketone (MEK), methylene chloride or m-pyrol (NMP). To prevent an explosive vapor mixture from being formed when the solvent is added to the reactor, the air must be replaced by nitrogen gas. The solvent needs to be heated to just above its boiling point and kept there until the solid material has been softened and removed from the metal. A second rinse with clean solvent may be needed.

As this process is an "on cost" to production, the cleaning must be done in a manner to keep down the costs while still maintaining quality. Dirty solvents can be reused until they will not clean efficiently. The solvents must be kept dry to prevent the dissolved prepolymer from coming out of solution.

All final waste material must be disposed of in accordance with local regulations.

4.3.10   Quality Control and Assurance

The process of producing a prepolymer is one of adding two or more liquids together under controlled conditions to obtain a prepolymer. To maintain the quality of the product, the following areas need to be addressed and controlled:

- Raw materials
- Process instrumentation
- Process conditions
- Final tests

The raw materials need to be of a suitably consistent quality and maintained at the same quality as it was received. Every effort must be made to prevent the ingress of moisture into any raw material. The moisture content of the major raw materials should be 0.05% or preferably less than 0.03%. Manufacturer certificates of analysis must be obtained for all materials purchased and the values used to make fine adjustments to the batches. In-house confirmation of quality of the raw materials may include:

- FTIR confirmation of identity of raw material
- Moisture analysis by the Karl Fischer method
- Viscosity, capillary method or spindle viscometer
- Hydroxyl number (for polyols)
- Isocyanate content (for isocyanates)
- Acidity (pH, 10% neutral alcoholic solution or titration)

With the widespread acceptance of the independently accessed ISO 9000 scheme, many companies will accept the analysis of the raw material supplier. Once the quality and consistency of the raw material is proven, only vital tests may be carried out. Due to moisture sensitivity, the opening and sampling of the drums may lead to faulty material if the drums are not properly resealed. The three vital control instruments are:

1. Weight
2. Temperature
3. Time

The weights of the raw materials added must be recorded either manually or by printout from the scales display unit. The weighing units must be regularly checked and recalibrated if needed. The capacity of the scale must be of a suitable size for the material being added. The low quantity of acid adjustment material needed requires the use of a lower total capacity scale.

If the liquid materials are being volumetrically pumped in, the amount must be correct for the temperature of the material to obtain the correct weight.

The basic position of the temperature probe will be determined in the construction stage of the reactor. The correct function of the actual probes must be checked on a regular basis. The actual temperature may be randomly checked using a noncontact infrared thermometer.

The time/temperatures obtained must either be recorded manually or logged automatically from the temperature controller. The point of any addition changes must be noted.

If the batch is produced on a manual system, more detailed readings and times need to be added to the batch sheet. With minimal equipment the weight and temperature instrumentation will provide either hard copy or computerized records of the time, temperature and weight of each batch.

If a batching controller is used, the program for each grade must be routinely verified for continued lack of corruption or unauthorized change.

Prepolymer production is a time-dependant operation that reaches an optimum point and then the properties are reduced. The production control must be such that the critical parameters of temperature and time are closely monitored.

The prime control is the viscosity of the material. This is best carried out using a temperature-controlled cone-and-plate viscometer. The sample volume should be small to allow for quick stabilization of the material to a temperature of 23 °C. A secondary test to determine the NCO content, a titration method based on that given in Appendix E, must be carried out as rapidly as possible.

The day after production, fuller tests can be carried out to confirm the quality of the batch. The tests must be carried out by approved local or international methods (i.e., ISO or ASTM). Final release of the prepolymer is based on these results and the data obtained during the production of the prepolymer.

4.3.11   Typical Process

The general procedure to produce a batch consists of several steps:

1. Prepare the batch card.
2. Program the batch controller if used.
3. Check that the reactor is clean.

4. Bring the raw materials to starting temperature (MDI may need to be melted).

5. Turn on the reactor heating and bring to the starting temperature.

6. Check all the valves for correct functioning.

7. Flush reactor with nitrogen.

8. Weigh in the isocyanate and record weight.

9. Bring the isocyanate to start temperature (40 to 50 °C).

10. Adjust acidity if required and record amount.

11. Slowly add polyol(s) over one to two hours.

12. If temperature rises too rapidly, slow down or stop polyol addition.

13. If temperature goes above 85 °C stop polyol addition and start external cooling.

14. After all the polyol is added, slowly ramp up the temperature to the end temperature and keep it at this level while the reaction comes to completion.

15. Apply vacuum to reduce isocyanate vapors and entrapped gasses.

16. Monitor viscosity and NCO levels over final period.

17. When the viscosity is correct (and NCO if tested), discharge the batch under slight positive pressure. Label batched material.

18. Proceed with second batch if required after reducing reactor temperature or else clean reactor and valves.

19. Test the prepolymer fully prior to release.

4.4 One-Shot System

With the use of suitable catalysts, it is possible to produce a polyurethane product in a single step. This involves mixing the correct ratios of the polyol(s), isocyanate, catalyst and any pigment and plasticizer.

A major issue is the exotherm that is given out in the reaction. The heat generated must be removed in the molding. The total amount of heat given out is the sum of the heat from the initial prepolymer reaction and the final chain extension stage. This leads the process to be used in relatively thin walled products.

The materials must be at the correct temperatures and the flow rates set correctly. The machine must be purged until the mix is constant and correct.

Figure 4.5 illustrates the ingredient flow in the one-shot method.

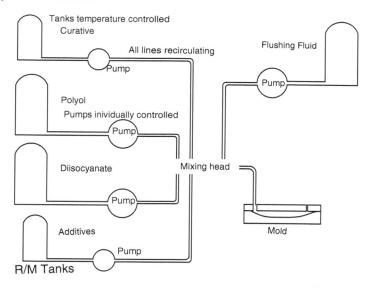

FIGURE 4.5

Ingredient flow in one-shot polyurethane processing.

4.5 Quasiprepolymers

Polyurethanes made using MDI are normally cured with a diol. The diols have a low molecular weight and are hygroscopic. Using manual systems, keeping the diol such as BDO dry is a difficult procedure. Overcoming this problem led to the development of quasiprepolymers. Quasiprepolymers are normally taken as having at least a fourfold molar excess of isocyanate in the isocyanate side of the system. The other side contains the remainder of the polyol, curative and any catalysts. A secondary advantage is that the instantaneous release of heat is reduced.

Polyurethane systems can be produced to have near-equal or easy mixing component ratios (either by weight or volume). The prepolymers in this type of system are called quasiprepolymers. The preparation of the isocyanate side is similar to that of a standard prepolymer, with the difference being that only some of the polyol is reacted with isocyanate. All handling procedures are the same as for the normal prepolymer except the application of a vacuum at the end of the reaction. Automated processing units such as used in the one-shot systems have become more affordable and more accurate. The handling of the BDO is also now much easier as it can be kept under a dry inert atmosphere. Commercially there are two types of systems on the market:

1. Two-part systems where the material when mixed in the specified ratios will produce a polyurethane of the desired hardness and properties

2. Three-part systems where the materials are mixed in varying ratios to produce polyurethanes with different hardnesses and properties

Two-part systems In these systems an exact balance can be obtained between the end properties and the raw materials used, for example the molecular weight of polyols and the mixing ratios of the polyols/diisocyanate/chain extender.

Three-part systems The major advantage of this method is that with a small range of raw materials a wide range of polyurethanes can be produced. In this method there are two standard sides, one containing the diols and the other a quasiprepolymer of 20 to 27% available isocyanate. The third is a chain extender such as 1,4-butane diol (BDO). Mixing the three components together in different ratios, a range of materials can be made from soft to hard (65 Shore A to 65 Shore D). This type of system makes it very usable in machine processing.

General As a portion of the isocyanate has already reacted, the exotherm is generally not as large as with the one-shot system. Viscosity and NCO checks need to be carried out.

In the United States, the isocyanate side is called the A side. If the material comes from Europe, the reverse is true and the material may be called the B side. Care must be taken and the appropriate checks made.

In the preparation of the polyol or B side of the two-part system, the curative is dissolved in the remainder of the polyol. Care must be taken that the curative (often BDO) is fully soluble in the remainder of the polyol at all temperatures. It is essential to examine the material over time to check if there is any phase separation. Manufacturers provide solubility data for BDO in various polyols but it is best to check your own system.

Depending on the temperature at which the material is processed, various amounts of a metallic catalyst are added. Tin-based catalysts like Air Products T12 catalyze the OH-NCO reactions. T12 increases the total reaction speed but also equals out the reaction rates of the high molecular weight polyols and the low molecular weight polyols. Bismuth catalysts are now preferred over the previously used mercury catalysts. Polyether polyols are slow in reaction and it may be necessary to use both a tin and an amine catalyst to obtain the best speed and properties.

References

[1] API. TDI Transportation guidelines. *Trade literature brochure*, pages 1–35 plus Appendices, 2002.

[2] BASF. Polyurethane MDI handbook. *Trade literature brochure*, pages 1–31, 2000.

[3] Dow Chemical Company. Isonate 125M -Pure MDI. *Trade literature brochure*, (026):1–4, 2001.

[4] Dow Chemical Company. Prepolymer Production. *Trade literature brochure*, pages 1–16, 1960.

[5] Huntsman polyurethanes. Ssuprasec TDI 80/20 toluene diisocyanate (80/20 isomer ratio). *Trade literature brochure*, (rev 8/00):1–2, 2000.

5

Processing Fundamentals

5.1 Introduction

5.1.1 Process Definition

The production of solid polyurethane parts of the correct engineering quality requires the conversion of either the prepolymer or quasiprepolymer to a solid material. The grade and chemistry of the material must be carefully considered in order to obtain a material that can be reproducibly processed and that has the correct final properties.

The correct application of heat must also be used to obtain the best product.

5.1.2 Importance of the Process

Curing, or more correctly chain extension, is required to convert the material from a semisolid or liquid into a solid material which possesses elastomeric or rubbery type properties. The definition of an elastomeric polyurethane is that on the application of a strain, the material will stretch and when the strain is removed it will rapidly return to approximately the original length. This increase in length should be less than 1%. In the case of fully cured polyurethane the material is normally very tough and the extension can only be made by a machine. In a thin sheet it may be illustrated by folding it double and then letting it snap back.

It can be considered that there are three main groups of polyurethanes:

1. Thermoplastic polyurethanes

2. Castable polyurethanes

3. Millable polyurethanes

Thermoplastic polyurethanes (TPUs) are normally processed in conventional plastic machines and when heated to above 120 to 150 °C will soften and can be processed. By definition this process can be repeated over and over again. The TPU is supplied as a polymer chain extended to a suitable length with terminal groups that do not allow any further chain extension.

Castable polyurethanes are supplied as a prepolymer with an active terminal isocyanate group to the polymer chain. These isocyanate groups are reacted with either a diamine or a diol. In the case of a quasiprepolymer, the prepolymer is made and the chain extension carried out at the same time. As the chain length becomes longer, the viscosity increases and at a certain point it becomes a solid. On further heating, extra strength is developed by a type of physical chemical bond (hydrogen bonding). In castable polyurethanes the actual chains will break down on heating before the physical chemical bonds give way. The material can therefore not be reformed after the chain extension is complete.

If a triol is used instead of a diol (a typical example is the addition of TMP), the resultant polyurethane is both cross-linked physically as well as by hydrogen bonding. This physical cross- linking leads to a breakdown of the hard segments. The density of the hard segment zones is a major factor in the hardness of the material. The use of triols will make the material much softer as it breaks up the hard segment zones to a certain degree. The material made is softer but has better compression set.

Millable polyurethanes are modified thermoplastic polyurethanes that contain special sites for cross-linking. They are normally processed using rubber processing equipment. Typically in this group of polyurethanes the chains are modified by inserting a chemical "double bond" in the soft segment. These can be cross-linked using a peroxide curative or by using a highly accelerated sulfur cure. Very good properties can be obtained.

5.1.3   Changes in Material Properties Before, During, and After Curing.

The process of curing polyurethane elastomers depends to a large degree on initially having a material with a reasonably low viscosity. This is to allow the flow of the material into the mold and the ability for any entrapped air to rise to the surface. Once the mold has been filled and the air allowed to escape, the increase in viscosity should be rapid to prevent excessive leakage through any parting lines in the mold.

Heat plays an important part in the curing of polyurethanes. The reaction itself gives out heat so this must be taken into account in determining the temperature of the mold. The mold should be at the maximum temperature that the curing prepolymer will reach. An MDI-based system will release heat more rapidly than a TDI-based system; therefore the mold has to be hotter than when using a TDI-based material.

It must also be noted that there is a certain amount of shrinkage due to the actual reaction. The harder the fully cured polyurethane, the more the final shrinkage will be.

During the curing process the material builds up a degree of strength. The strength after several hours is usually sufficient to allow the product to be removed from the mold without any damage to it. At this point full mechanical properties have not yet developed. Further heating of the material at 80 to 100 °C for 16 to 24 hours is required for all the properties to be fully obtained. The purist of the polyurethane processors will maintain that the polyurethanes should then be rested at room temperature for one to two weeks to obtain the ultimate in properties. The explanation of this is that the polyurethanes will slowly absorb water from the air that will assist the formation of very strong bonds.

5.1.4 Stages of Curing and Factors Affecting Cure

Prepolymer Curing or Chain Extension The prepolymer as received from the manufacturer has a simple chain that has been terminated with an isocyanate group. The isocyanate ends with this magical NCO group. The NCO is the reactive part of it. The higher the percentage of NCO in the prepolymer, the harder the material will be. An 80 Shore A will have an NCO of approximately 3.1 to 3.2%, whereas a 75 Shore D will have an NCO content of about 11.2%. To obtain the chain extension, one must add an appropriate amount of amine or diol curative. For every curative there is a different amount that must be added. The manufacturers of the prepolymers and curatives will give the appropriate factors for mixing the polyurethane. The prepolymer must be heated before use. This is to reduce the viscosity of the material as well as to obtain the correct cure rate and complete cure time.

The amine and the prepolymer each have two available sites for reaction. The amine will react with one isocyanate (NCO) and then the second amine will react with another isocyanate. This reaction will in effect double the length of the chain. The reaction will continue until all the amine has been used up. As the chains become longer they become more entangled and the viscosity of the material increases. At a certain point, some order begins to appear in the mixture and the hard segment areas all tend to agglomerate into a group and the soft or body section of the prepolymer remains separate. These agglomerated areas form hydrogen bonds and begin to give the material strength. Careful experimentation will show that after two to three hours, 95% of the available isocyanate groups have been used up (at 80 °C). After 14 to 16 hours, over 99% has been used up.

The final properties of the polyurethane can be enhanced by using slightly more or less than the theoretical amount of curative needed to react with all the available isocyanate groups. Table 5.1 details the changes in properties with respect to the level of curative.

Quasiprepolymers (Cold Cure Materials) One of the major advantages of quasiprepolymers is the ability to produce parts using a system with very even mixing ratios. The reduction in errors when using small quantities of highly hygroscopic materials such as BDO is very important. Unless

TABLE 5.1
Property Changes with Different Curative Levels

Property	Effect of Changes in Mixing Ratios
Physical Properties	
Hardness	Remains unchanged between 85 and 100% theory
Tensile strength at break	Maximum physical properties are achieved between 90 and 95% theory
Modulus	Stable; minor change over the range of 85 and 100%; decrease outside this range
Elongation	Maximum elongation at 100 and 105% theory
Tear strength	Maximum properties at 100 and 105% theory
Compression set	Best at 85 and 95% theory
Dynamic Properties	
Flex life	Maximum property at 100 and 105% theory
Abrasion resistance	Remains relatively unchanged between 85 and 105% theory although 100 and 105% is the best
Resilience	Maximizes at 85 and 90%; slight decrease above this range
Hysteresis, Dynamic Mechanical	Low percent stoichiometry is preferred
Other	
Heat resistance	Best at 85 and 95% theory

The theoretically calculated value of the amount of curative needed to react with all the available isocyanate is called the 100% theoretical or stoichiometric value.

prepared in-house, the disadvantage is that the chemistry is controlled by the manufacturer of the system.

When using these systems one must always ensure that both sides are individually very well mixed prior to final blending. The manufacturer of the system will make every effort to ensure that the components are all fully soluble in each other but on occasions they will separate out.

When these materials are processed, two reactions take place. The polyol side will react with the isocyanate and the curative will extend the polymer chains to their maximum length.

5.2 Introduction to Molding Process

The distinctions between advantages and limitations of batch and continuous processes are given in Table 5.2.

TABLE 5.2

Comparison between Batch and Continuous Processing

Batch	Continuous
Low capital cost	Cost of machine $40,000+
Able to change grades rapidly	Suitable for longs runs on one grade
Minor variations batch to batch possible	If machine running well, uniform quality
Large job requires many batches	Continuous pour (within limitation of machine)
Manual process	Process can be automated
Operators more exposed to chemicals	Can be completely self-contained

Continuous processing machines allow for the dispensing of many small items with the output being stopped for a short period in between each mold. The maximum size pour is dependent on the size of the feed tanks.

Hand casting of polyurethane elastomers is summarized in Figure 5.1 and more details are given in subsequent sections.

Before commencing any casting operations, the people involved must be fully conversant with the health and safety requirements of the operations that they are about to carry out. They must be familiar with the SDS details of all products being used as well as the engineering and personal protective equipment required.

There are seven main steps in hand casting polyurethane parts:

1. Prepare and mold-release mold.

2. Preheat prepolymer if required and degas.

3. Add pigment and any other additive to prepolymer.

4. Prepare curative.

5. Mix prepolymer and curative.

6. Pour into mold where curing will commence.

7. Trim article and fully cure molding.

1. **Prepare Mold**

   Molds must be checked for damage prior to use and preheated to a temperature equal to the maximum that the exotherm will reach when the

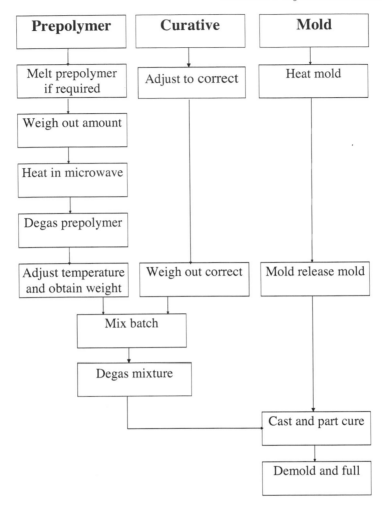

FIGURE 5.1
Hand casting of polyurethane elastomers.

prepolymer and curative are reacted together. If required, a thin coating of mold release is applied to the mold to assist in the demolding process. It is often found that more that one release can be obtained from the mold that has previously been prepared.

2. **Preheat Prepolymer**

The prepolymer must be melted so that all solid material is liquid. (A lump of unmelted prepolymer is sometimes seen in the middle of the drum). It must be remembered that excessive heating irreversibly ages the material. Once a drum or pail has been melted, it can be kept in that state by keeping it in a warm spot in the factory. The prepolymer can be weighed out and heated in a microwave or conventional oven to the required temperature. The material should be degassed under high vacuum prior to use. This helps to remove any gasses and moisture in the material. If the material is very foamy, breaking the vacuum during operation will help. Manufacturers will provide material that is both very dry and gas-free. For articles that are noncritical this step is often left out. The operator must ensure that it is safe to eliminate this step.

3. Add Pigment or Other Additive

If required, pigment and any other additive may be added and predispersed. The pigments used must be in a dry nonreactive medium. If the pigments are dispersed in a polyol, the paste may absorb moisture and will react with some of the isocyanate groups. Some yellow pigments need to be added to the prepolymer prior to its final heating because they disperse poorly.

4. Prepare Curative

Curatives may either be solid or liquid at the processing temperature. Solid curatives such as MOCA or M-CDEA are normally melted under controlled conditions prior to use. The manufacturer's recommendations must be taken into account as to the temperature the material is heated to. Excessive heating can cause breakdown of the materials, which is dangerous from a health and technical point of view. MOCA, if heated to 140 °C, will start to decompose and give off dangerous fumes. The molten curatives can cause severe burns to the skin and absorption into the bloodstream.

Solid curatives may be dissolved in a liquid solvent to eliminate the handling of the hot molten material. The solvent may be a nonreactive plasticizer such as a phthalate, for example DIOP, or an ester such as a benzoate, for exampl, Benzoflex 9-88. These will soften the product but not take part in any reaction. If a diol or triol of high molecular weight is used, it will take part in the curing reaction. The equivalent weight of the system must be taken into account when calculating the weight of curative required. In both cases the hardness as well as the physical properties will be changed.

Liquid curatives, in a number of cases, are very hygroscopic and great care needs to be taken to prevent the absorption of moisture vapor from the air. The material needs a constant blanket of nitrogen, and, if required, dried molecular sieves to keep the material dry.

5. Mix Prepolymer and Curative

Mixing tools used should be of dry impervious materials. Wood is not considered a good mixing tool material. The mode of mixing should be such that there is a minimum of air entrapped in the blend. The mixing could either be in a figure-eight or a zig-zag motion across the container. See Figure 5.4 later in the chapter. The sides and bottom of the container must be scraped to help remove the prepolymer from the sides of the wall and bottom of the container. There must be no swirl lines in the blend when the mixing is finished. These lines represent bad dispersion. If power mixing is used on large batches, the mixing head must always be immersed and the speed adjusted to prevent inclusion of an excess of air. If air is introduced into the mix and there is sufficient time, the mix may be degassed for a second time. The mixing technique must be adjusted to add the minimum amount of air that is possible.

6. **Cast into Mold** Once the prepolymer and catalyst are added together, the chain extension (curing) reactions will commence.

   The time taken to mix must be carefully monitored. It must be sufficient to allow complete mixing but there must be enough pot life left to allow pouring into the mold while the material is at the lowest viscosity possible. This is to allow the polyurethane to fill the mold completely and for any entrapped bubbles to reach the surface.

   Polyurethanes tend to trap air very easily. The method of pouring must be such that the mix will flow down a surface, across the bottom, and fill the mold upward, displacing the air ahead of it. The pouring should be in one continuous stream. It may be advantageous to raise the height slowly during the pour to improve the flow of the material into the mold.

   Any exposed bubbles may be gently popped using a soft gas flame or a hot air gun. The heat source must be used to a minimum and kept moving to prevent damage to the surface.

   When the material has gelled sufficiently for the mold to be moved, it should be placed in an oven heated to the curing temperature for the initial curing to take place. The reaction proceeds and the structure starts to develop. When the material has sufficient strength, the article can be demolded. MDI/BDO polyurethanes normally need more initial cure time to build up sufficient structure to allow for successful demolding.

7. **Trim Article and Fully Cure Molding in Oven** The state of cure at this point allows for the easy removal of a large proportion of any flash that may be present. This can be carried out using a sharp knife with the material under slight tension.

   The material at this point will appear to have sufficient properties for use but the full cure cycle is needed to bring about the complete cure. The article needs a further two weeks at ambient temperature to reach its absolute maximum properties.

   Post molding operations such as final cleanup, trimming to size or minor repairs can be carried out immediately after the full cure cycle.

5.3 Health and Safety Concerns during Casting

The casting operation will expose the personnel involved in close proximity to both chemical and physical hazards.

Copies of Safety Data Sheets for all the various chemicals used must be available at all times. These documents must not be more than three years old. They include all prepolymers, curatives, solvents, plasticizers, adhesive paints, blast cleaning agents, solid and liquid additives, and any gasses used.

Inserts that are to be used in the molding must be prepared by either grit blasting or chemical treatment. The grit blasting operations must be carried out in a sand blast cabinet or the operator must be in an approved air-supplied garment.

If solvent-based adhesives are used, appropriate personal protective equipment must be worn (chemical mask, gloves and apron) and the operation carried out in a fume extraction hood.

Molds can be made from either metallic or nonmetallic materials. Metallic and nonmetallic molds are heated to the same temperature which can be up to 120 °C. Metallic molds are very much more heat conductive than molds made from polyurethane or epoxy. When handling the metal molds, clean heat-resistant gloves must be worn to prevent burning of the skin. The weight of these molds must be taken into account when handling them. Suitable lifting equipment such as a crane, block and tackle, or a forklift must be used.

When the prepolymer is melted, it and any gas in the container will expand. Care must be taken to prevent the pressure from building up excessively. The container must be vented to reduce the pressure buildup; otherwise it may rupture. After use the container must be flushed with nitrogen to remove any moisture containing air.

Vessels used for the degassing of prepolymers must be designed to withstand the 250 to 280 Pa gauge vacuum applied. The exhaust fumes from the vacuum pump must either be trapped or exhausted safely because they will contain isocyanate vapors.

If silicas are being used as a filler, proper handling and mixing equipment must be employed as these materials are extremely fine. Both engineering and personal protective measures must be used.

MOCA is classed as a suspect carcinogen so its handling must be in straight accordance with the local safety rules and regulations. The manufacturer's recommendations must be observed for the use of all curatives.

During the casting process, care must be taken not to inhale any vapors from the mix (use suitable respirator) and to prevent any burns or absorption through the skin. All solvents must be removed before the flaming off is carried out to reduce the fire hazard risk. Both cast and cured polyurethane moldings are hot and may be giving off fumes. Heat, chemical and weight considerations must be taken into account when handling the filled molds.

The degree of difficulty in demolding the parts is a function of the design of the mold and the correct functioning of the mold release. Any levers used must be free of burs and cracks. Compressed air is often used to assist in the breaking of the seal between the polyurethane and the mold. The air lines and gun must always be in good working order. Accidental misuse of the compressed air is very dangerous.

5.4   Mold Preparation

5.4.1   Molds

Molds used in the processing of polyurethanes can be made from a variety of materials. These include:

- Aluminum

- Polyurethane

- Silicone

- Epoxy

- Hard thermoplastics

- Steel

The polymer molds may or may not be filled. The filling material can be a metal such as iron shot or silica particles.

The first consideration is to check that the mold is the correct one for the product that is planned to be made. The molds should be permanently identified.

Before using a new mold, the dimensions should be checked. The coefficient of expansion of the mold material must be balanced against shrinkage of the cast polyurethane (normally 1.0 to 1.5%). It is normally much easier to remove material from a mold or master piece than to add to it. This should be taken into account when designing the mold.

The first part produced from any new mold should be fully cured, allowed to cool and then fully checked for dimensions. Polyurethanes absorb moisture at high humidity and will expand. At ambient temperature and 100% humidity the expansion may be +0.6%.

The presence of any air pockets formed on the casting should be noted in relationship to their position in the mold. Additional vents may need to be added at this point if required.

5.4.2 Cleaning and Repair

Before reusing a mold it should be physically cleaned to remove any grime or protective materials. Buildup on the surface should be cleaned off without damaging the surface of the mold.

Location pins must be checked and repaired if needed. Any damage caused during demolding must also be repaired if in a critical area. The repairs must be such that they do not leave any marks on the surface of the part.

5.4.3 Mold Release

There are three main groups of mold releases namely:

1. One-shot

2. Semipermanent

3. Durable film

A selection of commercial mold release materials is listed in Appendix C.

All mold release agents should be applied in accordance with the manufacturer's instructions. Thin coats of the mold release should be used. Great care must be taken not to get any mold release onto any inserts that have been coated with a bonding agent. This will prevent the bonding of the urethane to the insert.

One-Shot Mold Release

This group is commonly based on silicones and needs to be applied on a regular basis. Initially the release agent needs to be used after every pour but often there is a buildup and several pours may be done between applications.

Water-based mold releases are available. Care must be taken with these materials to ensure that the water used to carry the release agent is completely evaporated before use and the mold brought up to the correct operating temperature.

Semipermanent

These materials can provide many releases from a mold (20+) before recoating is required. They often require more preparation before use such as baking on. The agents are often based on fluorine polymers.

Durable Film

These films can be based on Teflon® [1] film, polypropylene or Mylar® polyester film. These release materials are normally used where the surface of the material must be free of any contamination and defects. An example of the application is in the preparation of tests sheets for determining mechanical properties.

---

[1] DuPont, Wilmington, Delaware.

5.4.4   Assembly

After the mold has had a coat of mold release applied, it must be fully assembled. Inserts must be correctly placed in the mold and fully secured. In a complex mold it may be necessary to temporarily lock the insert into place with a removable pin that is taken out after the mold is full but before the material has gelled. Any join lines that may be prone to leakage must be sealed with a silicone sealant.

5.5   Batch Size Adjustment

5.5.1   Quantity of Polyurethane

If the mold has been made from a CAD/CAM design, there is often a facility in the package to estimate the volume of material to use. The weights of the smaller and larger parts in a series can be used for gauging the amount of material required.

Sufficient material to overfill the mold should be prepared for the first pour so that the mold will definitely be filled. Noncritical molds should be available to use up any excess material prepared.

5.5.2   Weight Calculations

The next step in the process is to calculate the quantity needed to make the articles. The main factors to be considered are:

- Volume of the mold

- Density of the final polyurethane

- Formulation of the material

A simple example is given below:

Prepolymer	100.0
Curative	10.0
Pigment	0.5
Total:	110.5

The density of the polyurethane must be known. This can be determined experimentally or the material supplier's data used. In this example the volume is assumed to be 1.10 liters and the density of the polyurethane is 1.07 g/cm^3.

The most important point is that all the data must be in the same units. After doing the calculation, make sure the result you get is logical.

$$1.10 \text{ liters} = 1100 \text{ cm}^3$$
$$\text{Density} = \text{Weight/Volume}$$
$$\text{Hence, Weight} = \text{Volume} \times \text{Density}$$
$$= 1100 \times 1.07$$
$$= 1177 \text{ g}$$

Always allow for some material to remain on the inside of the container. A starting point is 5%. This should be cut back as experience is obtained. The adjusted batch size will then be:

$$(1177 \times 105)/100 = 1236 \text{ g } (1240 \text{ g})$$

Calculate the quantities of materials required:

$$\text{Prepolymer resin} = (100/110.5) \times 1236$$
$$= 1119 \text{ g } (1120)$$
$$\text{Curative} = (10.0/110.5) \times 1236$$
$$= 111.9 \text{ g } (112)$$
$$\text{Pigment} = (0.5/110.5) \times 1236$$
$$= 5.59 \text{ g } (5.6)$$

5.6 Prepolymer

Polyurethane prepolymers are supplied in a variety of different sizes, from small 1-kg cans up to 1-ton IBC containers. The available isocyanate (%NCO) level should be marked on every container of conventional prepolymer. Quasiprepolymers do not have individual batch levels supplied as the overall reaction levels are factory adjusted.

The level can be confirmed by a standard dibutyl amine titration. A typical method is given in Appendix E. Depending on the type of isocyanate used, the exact solvent for dissolution of the sample may have to be determined.

The prepolymer should be used on a first-in first-out (FIFO) basis as it may deteriorate slowly on standing. During the normal manufacturing process the system is stabilized to reduce the aging process. Systems based on the unstable prepolymers (such as NDI-based materials) only have a life of several hours so this must be taken into account.

Once a container has been opened, care must be taken to prevent moisture from the air affecting the material. This may be simply done by replacing any air in the container with dry nitrogen gas prior to resealing or in a more permanent situation, keeping a supply of nitrogen gas over the prepolymer. The pressure in the drum cannot be allowed to build up when being heated as this may cause the possible rupture of the drum. Industrial-grade nitrogen with a dew point of −40 to −50 °C is sufficiently dry for this purpose.

The threads of the bungs must be cleaned after material has been decanted from the container to prevent sealing of the bung. If the bung cannot be opened, the gentle application of heat around the bung is often sufficient to soften the partially cured polyurethane and allow reopening of the container. Excessive heat must not be applied and any seals must be checked after this method is used.

Polyurethane prepolymers may be liquid or semisolid at ambient temperatures. Before further processing can take place they must be heated so that the material is fully melted and the viscosity reduced to a point that it can be dispensed successfully.

The container may be heated in a thermostatically controlled hot room or with a heating band. Care must be taken to prevent localized overheating. As the thermal conductivity of the material is poor, there is often a temperature gradient in the container leading to the outside being melted and the center being still solid. This can be corrected by roll-mixing the container. All the material needs to be molten prior to use. The can must not have any translucent lumps in it.

This melting of the material adds to the heat history of the prepolymer. Table 5.3 is a general guide to the heating that a typical prepolymer can withstand.

TABLE 5.3
Prepolymer Life

Temperature, °C	Time
25	1 year
60	2 weeks
70	3 days
80	36 hours
90	12 hours
100	8 hours

The polyurethane prepolymer will slowly increase in viscosity on prolonged heating. It is a result of a chemical reaction between the polyurethane chains. The chemical reaction is nonreversible. The available isocyanate is used up and therefore the stated %NCO is reduced.

Small volumes can be heated in a microwave oven. Amounts of polyurethane too large for the microwave oven will need to be heated in a forced draft oven. The prepolymer must be heated to slightly above the required temperature. The required temperature depends on the prepolymer and curative system being used. The basic data is normally supplied by the

TABLE 5.4
Typical Supplier Data

Polymer Type	Polyester Based	Polyester Based	Polyether Based	Polyether based
Prepolymer level	100	100	100	100
%NCO	3.3	5.5	4.15	6.3
Curative	MOCA	MOCA	MOCA	MOCA
Curative %, theory	95	95	90	90
Curative level	10.0	16.6	11.9	18.1
Processing temp.:				
Prepolymer, °C	93	82	93	83/93
Curative, °C	113	113	113	113
Pot life, minutes	8	4	10	5.5

manufacturer of the prepolymer or curative. Typical data is shown in Table 5.4.

The larger the part, the lower the temperature the material must be processed at. This is due to the fact that as the material cures, heat is given off. In the center of thick cross-sections the heat cannot escape as the thermal conductivity of the material is poor. Undesirable reactions may take place at these temperatures. There will also be thermal stresses induced into the molding.

A balance must be reached between having a low viscosity to allow easy pouring and the speed of reaction. There may be other materials available that will give a lower viscosity while still having the required properties. The changes in viscosity with heat are shown in Figure 5.2.

To determine the quantity of curative that must be used, there are three main variables in the calculations:

1. NCO level of prepolymer

2. Type of curative used

3. Index value

When prepolymers are manufactured they do not always come out exactly the same. The viscosity and the isocyanate level can vary within the specified limits. The manufacturer determines the exact level of NCO that is in that batch and supplies that information on the container.

Each type of curative on the market has a different factor (molecular weight/functionality). This value changes according to the basic chemical composition and the fundamental activity of the curative; for example,

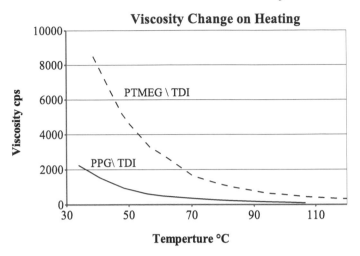

FIGURE 5.2
Short-term change in viscosity on heating prepolymers.

Material	Equivalent Weight (EW)
MOCA	133.5
Ethacure E300	107.8
BDO	45.0

The term "equivalent weight" (EW) is given by:

$$\text{Equivalent weight (E/W)} = \frac{\text{Molecular weight}}{\text{Functionality}}$$

The functionality is the number of appropriate active sites in the molecule.

Index value (% theory): By experience and research it has been found that to obtain the best results for an application, slightly more or less than the theoretical amount of curative should be used. The value is obtained by evaluating the most important physical properties required for the application in which the product will be used. This is discussed further in Chapter 8 Section 8.1.4.

5.6.1   Calculations

There are different ways of doing the calculations. Most commonly used methods are the %NCO method and the amine equivalent method.

%NCO Method

The calculation is to determine the amount of curative needed for 100 parts by weight of the prepolymer.

$$\text{Curative per 100 parts prepolymer} = \frac{\%\text{NCO} \times \text{EW Curative} \times \% \text{ Theory}}{\text{EW NCO} \times 100}$$

where

%NCO is the amount of terminal isocyanate in the prepolymer and is supplied by the manufacturer and is marked on the can

EW curative is supplied by the manufacturer of the curative

EW NCO is the equivalent weight of the isocyanate group and is 42.02

% Theory is the variation required to give the best properties for the system

EW of the curative is as given earlier and a fuller list is given in Appendix B.

The desired index is normally in the range of 80 to 115%. Normally the index value is between 90 and 105%. As there are several semiconstants, the calculation can be simplified.

The formula is normally used in a simplified form. For example, if we are using MOCA as the curing agent and the index at 95%,

$$\text{Curative per 100 g prepolymer} = \frac{\%\text{NCO} \times 135.5 \times 95}{42.02 \times 100}$$
$$= \%\text{NCO of prepolymer} \times 3.18 \times 0.95$$
$$= \%\text{NCO of prepolymer} \times 3.02$$

Amine Equivalent Method

The other method that is sometimes proposed is the "amine equivalent" (AE) method:

$$\text{Parts curative} = \frac{(\text{Curative EW}) \times (\% \text{ Theory})}{\text{prepolymer AE}}$$

The amine equivalent is given by:

$$\text{Amine equivalent} = \frac{\text{Formula weight NCO}}{\%\text{NCO}}$$
$$= \frac{42.02 \times 100}{\%\text{NCO}}$$

Manufacturers of prepolymers will often provide the weight of curative per 100 units of prepolymer. These are based on the midpoint NCO for the grade and the normal index value for the curative.

The calculations can readily be carried out using a calculator or spreadsheet. From a practical point of view a set of shop scales with a discount function has been used to provide the weight required.

After dispensing a run of prepolymer from a container, it must be resealed in a state to prevent attack by moisture in the air. A blanket of dry nitrogen gas must be used to displace any air present in the container. Any bungs or taps must be properly cleaned using a dry solvent such as MEK or MIBK to prevent traces of polyurethane from reacting with the moisture in the air and sealing the container.

5.7   Pigments and Additives

Pigments are used to color polyurethanes, basically for appearance. Normally cured MDI- and TDI-based polyurethanes yellow on exposure to ultraviolet light. The product will change color, becoming more yellow and darker where the light strikes the object. White or pale-colored pigments in polyurethane are not normally recommended for this reason.

Raw pigments will have to be ground in a dispersing medium to develop the full color of the pigment. The dispersing medium is normally a nonreactive plasticizer that is both dry and nonhygroscopic. The cost of a pigment system is a function of both the light and heat stability of the particular pigment and its tinting strength. The light and heat resistance are normally given as a ranking out of five. A ranking of 1/5 is very poor and 5/5 is excellent.

The chemical nature of the pigment must also be considered when using it in a system that requires FDA or equivalent approval. Cadmium- and chromium-based pigments are often suspect.

The quantity of pigment used should be kept to a minimum, normally in the range of 0.25 to a maximum of 5%. The pigmenting effect will reach a maximum and then stay relatively constant. Too much pigment will only degrade the quality of the part and add extra cost.

The ability of the pigment system must be checked prior to full production. Some pigments do not disperse well if introduced late in the mixing. Some yellows need to be predispersed prior to final heating for the maximum effect. Poorly dispersed pigments will appear either as flakes in the system or as swirls. The flaky material should be dispersed in the prepolymer before the final heating. Swirl marks normally indicate a lack of proper mixing although some marks are very difficult to remove in certain red pigments.

Pigments can be dispensed using plastic squeeze bottles with long nozzles. If the viscosity of the material is too high it may be diluted with some plasticizer. If larger amounts need to be used, tinting machine dispensers are an option.

Plasticizers can perform several roles in a polyurethane compound. There are two major groups of plasticizers used. The most common are the nonreactive plasticizers such as the ester group, and the second are long-chained-diols that act as a combined curative and plasticizer. In the case of the reactive material, this must be taken into account when calculating the amount of curative to be used (see Appendix B: Polyurethane curatives).

Nonreactive plasticizers will soften the polyurethane as well as extend the polyurethane's pot life. The mixture has a reduced viscosity, which assists in the ease of flow of the mixture and the removal of any entrapped air. This is of great importance in the casting of rolls where the final product must be bubble-free.

The use of fire-retardant plasticizers such as Fyrol^{TM}PCF at levels of 5 to 10 parts, will reduce the hardness by 3 to 4 points while providing improved fire resistance.

5.8 Curatives

Curatives are used to enlarge the prepolymer chain either by extension or by cross-linking. Materials with two active groups will extend the length of the chain by joining two chains together. Triols will provide the means to form permanent covalent bonds between different chains.

There are two main groups of curatives that are used. These are either hydroxyl- or amine-terminated compounds.

Table 5.5 gives a selection of some of the more common curatives. See Appendix B for a more comprehensive list. If a blend of curatives is used,

TABLE 5.5

Common Curatives

Type	Name	Equivalent Wt.
Diol	BDO	45
High M/W Diols	PPG 1000	500
	PTMEG 1000	500
Triols	TIPA	63.7
	TMP	44.7
Diamine	Ethacure E100	89.3
	Ethacure E300	107
	M-CDEA	187
	MOCA (MBOCA)	133

the equivalent weight of the blend must be calculated taking into account the individual equivalent weights.

$$\text{EW of blend} = \frac{\text{Wt. of A} + \text{Wt. of B}}{\dfrac{\text{Wt. of A}}{\text{EW of A}} + \dfrac{\text{Wt. of B}}{\text{EW of B}}}$$

For example, if a mixture of
3 parts TMP is used with 1 part of TIPA,

$$\text{EW of blend} = \frac{3+1}{\dfrac{3}{44.7} + \dfrac{1}{63.7}}$$
$$= 48.3$$

The above calculation gives a working EW of 48.3.

Certain amine-based curatives such as MOCA are classed as suspect carcinogens. Extra care must be taken to observe all safety requirements, both as required by local regulations and as specified on the SDSs. In all cases the requirements of the SDSs must be obeyed.

Curatives can be either liquid or solid. The solid curatives such as MOCA and M-CDEA must be melted prior to use. The prepolymer and melted curative must be at a temperature that will prevent any crystallization of the curative when added to the prepolymer.

The supplier's recommendations should be noted so that an adequate pour time is obtained. Pot life times quoted on the specification sheet are in many cases determined on laboratory-size samples where the effect of the bulk of material is less than in a large mix. In a large mix the poor conductivity of the material will cause the center of the mix to become hotter than the material on the side of the container.

Overheating the melted curative will result in partial decomposition of the material, giving off dangerous fumes. Care must be taken to keep the temperature stable and only to melt sufficient curative needed for the current run.

Protective clothing as recommended by the material supplier must be worn when handling molten curatives. The gloves must be impervious to the liquid curative. There is often a conflict at this point as heat-resistant gloves are also needed for handling the mold. Curative entrapped in cotton-style gloves can be absorbed into the skin.

Hydroxyl materials (diols and triols) are very hygroscopic and the absorption of moisture must be prevented. Positive nitrogen blanketing is often required. All air must be displaced by dry nitrogen gas and the material stored under a slight positive nitrogen pressure. The use of dried 5A molecular sieve in BDO is also advantageous.

5.9 Degassing

There are two main points on the processing of polyurethane where vacuum degassing may be required. The first is the prepolymer as supplied by the manufacturer. The second is after mixing the prepolymer with the rest of the system (curative, pigment, extenders and fillers).

The manufacturer of the prepolymer endeavors to supply a low-moisture, low free isocyanate, gas-free material. Moisture may be introduced with repeated opening of the prepolymer container. Contaminates will form bubbles in the final material. Any moisture or free isocyanates will have an effect on the reaction ratios as well as changing the chemistry to a degree.

There are two main methods to "degas" the prepolymer. The first is a simple batch system comprised of a vacuum pump and a pot designed to withstand vacuum. Figure 5.3 illustrates a simple vacuum pot. The alternate is a thin-film evaporator; in this system the heated prepolymer is allowed to flow in a thin film in a vacuum. Under these conditions the volatile material will be readily removed.

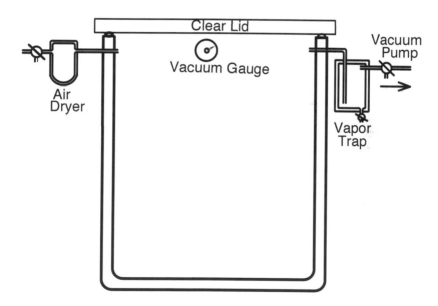

FIGURE 5.3
Vacuum degassing pot.

For safety reasons the pot must be specifically designed for vacuum use and any replacements be made from the same materials and the same size and design. A simple replaceable liner such as plastic wrapping in the pot will help keep the system in good condition.

The prepolymer is normally heated 5 to 10 °C higher than the final temperature. The volume of prepolymer in the mixing container should be approximately 25 to 30% of the total volume, as the material tends to foam as the volatile components escape. The viscosity of the material will drop as the temperature is increased. The lower viscosity and reduction in the boiling points of the materials under vacuum will allow the volatiles to escape more easily. A suitable silicone defoamer (Dow Corning® 200 Fluid) may be used if known not to give problems later in the process. These problems may include knitting, bonding to metal or creating a surface film that interferes with coatings. A few drops of a nonreactive dry solvent such as MEK may also be used.

The degassing should be continued until vigorous foaming ceases. There will always be some small bubbles showing in the polymer. The foam can be broken and knocked down by breaking the vacuum during the process. The inflowing gas should be dry. The material normally takes approximately 5 to 10 minutes at a vacuum of 250 to 280 Pa. The exact time will depend on the quantity, temperature, quality, degree of vacuum and shape of container used. The time needs to be determined experimentally for each set of conditions.

A second degassing may be carried out after the addition of all the ingredients to remove any bubbles introduced into the mix. It must be remembered that this takes up part of the available casting time.

5.10   Mixing and Casting

5.10.1   Premixing

It is most important that all materials must be dry before any weighing and mixing are carried out. As polyols may be rather hygroscopic, they should not be dried in an oven as they can absorb moisture from the oven. Fillers such as very fine amorphous silica have a very large surface area that can hold a large quantity of moisture.

Pours requiring more than one container must be carefully preplanned. The material must all be weighed out into separate containers and heated to the correct temperature. The auxiliary materials must be added to the correct containers and preblended. In large pours it is very important to add the next addition before the previous one has gelled. If the material has gelled too far, a knit line will form with consequent poor mechanical properties. Some time must be allowed, however, for entrapped air to escape.

5.10.2 Curatives

Plasticizers and catalysts can be mixed into the liquid curative prior to addition to the prepolymer if required. They must be well blended so that there is a uniform mix. The correct weight of the curative or blend must be accurately weighed in a clean dry container or with great care directly into the prepolymer. The material must be added in a thin continuous stream that will form a single phase. The addition must be done rapidly.

5.10.3 Mixing

Before the commencement of the final mixing, a last check of the mold must be made and the following points verified:

- Inserts prepared and seated properly
- Mold at correct temperature
- Thin film of mold release agent applied
- All joint lines properly sealed
- Pouring hole and vents open

When hand mixing is used, the stirrer must be made from a nonreactive material and of suitable size for the mixing container. The stirrer should have a straight edge that can scrape the sides and bottom of the vessel. Wood is generally not suited as a stirrer.

The motion of the mixing can be of any pattern as long as the material is completely mixed. Care must be taken not to whisk air into the mixture. There are two main patterns used, namely the figure-of-eight and the zig-zag patterns. The sides, bottom and corner must be scraped as prepolymer tends to hang onto the container. These two mixing patterns are illustrated in Figure 5.4.

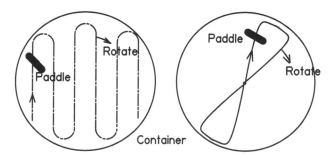

FIGURE 5.4
Hand mixing patterns.

The mixing must be complete with no signs of color streaks in the mix. Inexperienced operators should initially do mixes without any pigmentation. The mix in the container should be completely uniform with no lines. After pouring, the remaining material should be cured off. When the cured dregs are removed, there must be no sign of wetness. Any wetness is an indication of incomplete mixing.

If a mechanical mixer is used, a "Zippy" style head (such as used in the fiberglass industry) has been found suitable. The speed of mixing is important and the formation of a vortex must be prevented. Care must be taken with this style of mixer that air is not whipped into it. If necessary the mixture can be degassed for a short period at this time.

The time available for mixing is a function of the system and can vary from less than thirty seconds to four to five minutes. Once the prepolymer and curative are added, a chemical reaction commences. In very hard materials the time available is very short. After mixing, the material must still be poured into the mold and any entrapped air is allowed to escape.

5.10.4   Casting

Time

The mixing and degassing must be completed in sufficient time to allow complete filling of the mold and for any air bubbles to rise to the surface. The hotter the material, the lower the initial viscosity will be. However, the viscosity will increase more rapidly and the material will gel off more quickly. This is shown in Figure 5.5.

Pouring

The mixed polyurethane must be poured in a continuous stream from as low a height as possible. The polyurethane must be allowed to flow down a surface to the bottom of the mold. The bottom of the mold must be fully wetted by the mixture before the mold is filled from the bottom up sweeping any air out before it. Polyurethane will fold over on itself, entrapping air very readily. The pour hole should be two to three times the area of the vent holes. The mold may have to be held at an angle to allow a top corner to be fully filled by the material. The remaining material is added as the mold is slowly brought to the vertical.

The mold should be completely filled before the material has gelled. During the period that the polyurethane is still fluid, any air bubbles that have been introduced will slowly rise to the surface. The bubbles can be readily removed by briefly passing a soft butane gas flame over the surface to pop the bubbles. The flame must be kept moving and not allowed to degrade the surface of the material.

The onset of gelation can readily be checked by trying to draw a thread of material from the surface with a piece of wire about the thickness of a paper

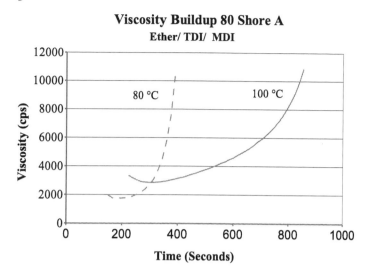

FIGURE 5.5
Viscosity buildups during initial curing of prepolymer.

clip. When a thread can be drawn, a lid can be placed on the mold or the mold transferred to an oven for curing.

5.11   Curing and Post Curing

On mixing all the ingredients, chemical reactions commence and the material will become more and more viscous and eventually solidify. The physical properties of the material are not fully developed at this point and the polyurethane will break very easily.

The mold containing the polyurethane should be placed in an oven that is set to a temperature 5 to 10 °C higher than the casting temperature. The ideal temperature is the maximum the molding will reach. This will reduce the shrinkage in the mold.

The casting is kept in the oven long enough to develop sufficient strength to allow the part to be successfully demolded without any damage or distortion. Careful removal of risers and sprues may be carried out at this point.

It is most important to realize that at this time the full properties of the material are not yet developed and the molded polyurethane must be returned to the curing oven for the period recommended by the system supplier. The curing time varies depending on the chemistry of the system.

The maximum properties of the cast polyurethanes are developed after a further one to two weeks at ambient temperature.

Polyurethanes with a hardness of 60 Shore D and above will often indicate full hardness but are very brittle. Extra post-curing cycles of ambient temperature to 120 °C will develop the toughness of the material.

6

Processing

6.1 Molding Methods

Apart from the simple open mold casting method, there are several more methods of casting polyurethanes. They include:

- Rotational casting
- Centrifugal casting
- Vacuum casting
- Compression molding
- Liquid injection
- Complex shapes

6.1.1 Rotational Casting

In rotational casting the heated mold is rotated in one or more planes at the same time. The viscosity buildup of the polyurethane is very important. In normal casting it is desirable to have the mix at a low viscosity for as long as possible before a rapid gelation. In rotational molding the viscosity needs to build up continuously at a steady rate. The pot life needs to be as long as practicable and in the order of eight to twelve minutes to allow for full and even coverage of the mold surface. Figure 6.1 illustrates the type of viscosity buildup required.

Flat sheets can be made using a rotating cylinder with lips on the ends. The cylinder must be level to ensure even thickness. The polyurethane mixture is added to the bottom of the horizontal drum, which is rotated at a speed of between two to fifteen revolutions per minute. The thickness of the sheet is controlled by the weight of polyurethane added. The speed of rotation must be adjusted to suit the diameter of the cylinder and the viscosity of the mixture. The cylinder must be fully coated before gelation starts.

When the material has built up sufficient strength, the casting can be removed and a sheet produced by cutting along the cylinder's length. The sheet is then allowed to fully cure on a flat surface in an oven.

This method allows sheets of even thickness to be produced without the need for a large flat heated table that must be level in both horizontal planes.

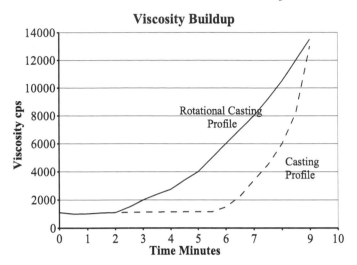

FIGURE 6.1
Viscosity buildup for rotational casting.

Different shapes including hollow spheres can be produced using this method. The major point is to obtain the correct rotation in the x, y and z planes.

6.1.2   Centrifugal Casting

Centrifugal casting is used to produce parts with fine detail relatively free from included air. The mold is attached to a horizontal spinning disk that rotates at a speed of 5 to 30 m/s. The mold must be capable of having the polyurethane added from the center of the disk. The centrifugal forces move the denser polyurethane to the circumference of the mold and the air moves to the center. Some air vents may be required as air can be trapped on horizontal surfaces.

The material needs to be added at a steady rate to allow it to flow evenly to the outside of the spinning mold. The mold, when full and the material gelled, may be removed and placed in an oven for the initial cure before demolding.

Warning

It is important to note that the mold must be dynamically balanced and securely locked in place on the spinning plate. The safety guards around the equipment must be capable of retaining the mold if it comes loose from the spinning plate.

The polyurethane mixture must be carefully added to the center of the mold. If it hits the edge of the inlet, strands of "fairy floss-like" polyurethane may be formed.

6.1.3 Vacuum Casting

In parts where amounts of entrapped air must be kept to an absolute minimum, the process may be carried out under full vacuum.

The unit consists of two chambers, one for the mixing of the prepolymer, curative and other ingredients, and the lower chamber for the mold. The mixing is done under vacuum and the fully degassed mix is poured into the mold using remote handling equipment. As both chambers are under vacuum, there is no air to be displaced and the mold is filled completely with no entrapped gasses. In certain units the mix can be pressure transferred to the mold. A pot life of at least five to six minutes is needed for this method.

The equipment is expensive and needs careful maintenance to keep the vacuum seals in good order.

6.1.4 Compression Molding

In this process the polyurethane mix is poured into a mold that can be placed into a compression molding press with heated platens. The material is allowed to gel and a top plate is placed on the material. The molding press is fully closed. The pressure applied is based on using 1.50 to 2.0 MPa (250 – 300 psi). The applied pressure is calculated using the projected area of the molding; for example, a part having a plan area of 300 x 400 mm in size:

Its area is 0.300×0.400 m^2 $= 0.12$ m^2

A pressure of $1.500 \times 0.12 = 0.18$ MPa on the mold is required.

The molds must be capable of withstanding the direct heat and pressure of the press. Aluminum or steel molds are preferred for this process.

Air vents are not strictly required in this process but it may be advantageous to briefly release the pressure and then reapply it to the molding. The exact time to apply the pressure varies from system to system and must be determined by experimentation.

A variation in the compression molding technique is to use transfer molding. In this method the liquid polyurethane mix is placed in a pot and allowed to thicken. The polyurethane semi-gel is forced through sprues into the mold and the air is allowed to escape through vent holes. The part is allowed to cure sufficiently so that it can be removed. Figures 6.2 and 6.3 show the general layout of a compression and transfer mold.

There are several key points in these methods to allow proper success. The pot, and if possible the sprues, must also be mold released to ease demolding. The sprues must be designed so that the remaining pad in the pot can be cut

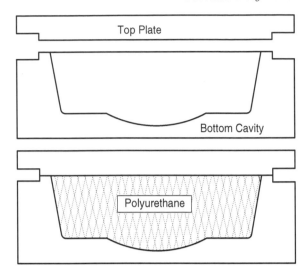

FIGURE 6.2
Compression molds.

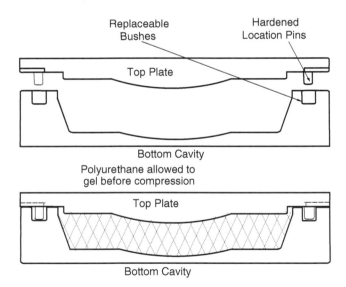

FIGURE 6.3
Compression molds more complex shape.

off. The taper of the sprues must be such that they are wider on the part side than the pot side. It is normally best to trim the sprues and material in the vents straight after demolding to ease the later trimming and cleaning of the part.

6.1.5   Liquid Injection

In this method the polyurethane is assisted in the flow by the application of gas pressure to the surface of the material. Parts where there is a long flow are particularly suited to this method.

The air ahead of the flow of polyurethane must be removed and other vents in the parts must also be used. Applying a vacuum ahead of the polyurethane will also help the flow and the prevention of trapped air. Typically pressures of 0.4 to 0.8 MPa are needed. The lower the viscosity, the lower the pressure required (Figure 6.4).

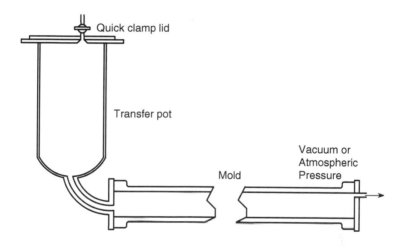

FIGURE 6.4
Assisted flow in molding.

6.1.6   Complex Shapes

Tall items with complex internal shapes that can easily entrap air are best filled from the bottom up. By introducing the polyurethane at the bottom of the mold, the polyurethane mix will rise, sweeping the air out ahead of the polyurethane. The easiest method is to have a removable delivery tube that reaches the bottom of the mold. When the polyurethane mix has filled the mold, the tube is removed before the material starts to gel.

6.2 Bonding

Polyurethane can be bonded to most materials either during the initial casting or after full curing. In both cases the most important consideration is that the surface must be clean and dry without any surface contamination.

6.2.1 Precasting Preparation

The polyurethane is normally bonded to a metal reinforcing or a rim of a wheel or gear. All bonding consists of four basic steps:

1. Removal of gross contamination
2. Preparation of surface (mechanically or chemically)
3. Application of bonding primer
4. Casting and curing of polyurethane

Cleaning Components

Reinforcings may be coated with a protective layer of grease that must be removed by an environmentally suitable method either using a locally approved solvent or a chemical alkaline wash. Gross scale and welding scatter must be mechanically removed.

Old polyurethane on rims may be removed either on a lathe, solvent attack or by freezing in liquid nitrogen. For more complex shapes, the reinforcing may be recovered by the above methods as well as by pyrolysis in a specially designed chamber where the material is heated in the absence of air to above the decomposition temperature of the polyurethane. The fumes are then burned using special after-burners.

A metallurgist should be consulted regarding both the freezing and the pyrolysis method as the reinforcing may pass through phase change temperatures. This may change the properties of the metal.

Surface Preparation

The surface of the metal must be treated to form a layer to which the primer can successfully adhere. The treatment may either be mechanical or chemical. Mechanical treatment is the most common and is either by grit blasting or hand abrasion. The grit or shot used must be clean and dry and free of any grease or oil. The blasting of the surface must be carried out in an approved grit blast chamber by a suitably protected operator. The operator must wear appropriate protective clothing to prevent inhalation of the fumes and dust developed during the process.

The surface of the part, when the blasting is complete, must be a matte gray with all the surface contamination just removed (ISO 8501-1 grade Sa2.5).

It need not be over-rough. The surface must be cleaned of excess dust from the operation. If needed, the surface may be cleaned with a solvent such as toluene or MEK. Solvents that evaporate too fast may, under some conditions, cool the part and cause condensation to form. The primer should be reapplied before reoxidation and contamination can take place.

The blasting medium used must be compatible with the material being prepared. The potential for chemical corrosion must be avoided. Copper slag should be avoided as it can react with steel and aluminum to form chemical cells later when the polyurethane absorbs moisture if standing or immersed in a liquid. Steel shot must also be avoided on aluminum. Aluminum oxide is a popular choice as the blasting medium.

6.2.2   Chemical Treatment

To obtain optimum adhesion onto a wide range of materials, chemical treatment of the surface is often recommended. The first action, as with the mechanical treatment, is to remove surface contamination such as grease with a clean solvent. The surface of the material is then modified to provide a layer for the primer to bond to.

The chemical treatment varies greatly with the type of surface. Some typical methods include the following:

Substrate	Treatment
Steel	The surface is subjected to phosphate treatment
Stainless steel	Dichromate treatment
Cast iron	Solvent and abrasion, carbon nodules in SG cast iron is a problem
Magnesium and its alloys	Dilute silicate/phosphate rinse followed by chromium treatment
Nickel	Nitric acid passivating
Titanium and its alloys	Chromic acid treatment
Zinc and galvanized metals	Dilute hydrochloric acid treatment
Concrete	Hydrochloric acid etch
Glass	Supplier's special primer

These chemicals are dangerous and full protective equipment must be used when carrying out the treatments. The appropriate SDS must be consulted prior to use.

Immediately after these chemical treatments, all excess treatment solution must be washed off and the part well dried.

The supplier of the primer should be consulted for current formulations to carry out chemical surface preparations.

6.2.3   Bonding Primer

There are specialist primers on the market for bonding polyurethane to metals. The three most popular brands are:

1. Chemlok® (Lord Chemical Products, Erie, Pennsylvania)
2. Conap® (Cytec Conap, Olean, New York)
3. Thixon® (Rohm and Hass, Philadelphia, Pennsylvania)

The prepared metal surface is coated with one or two coats of the primer, depending on the grade and the chemical and water resistance required. The coats must be applied at the recommended thickness. There is normally a minimum and maximum thickness. The aim is to fully wet the metal surface but not have too thick a layer. All the solvents in the primer must be evaporated before use. If the coated metal part is not used immediately, it must be carefully covered in polythene film to prevent surface contamination.

The primer can be applied either by brush, roller and doctor blade, or by spraying. To lower the viscosity for spraying, the recommended solvent must be used. The most important point when applying the primer by spray is to ensure that the spray is still solvent-wet when it reaches the part. Dry material can cause poor bond strength.

Once a part has been coated with a primer, it must only be handled with clean cotton gloves, preferably in a noncritical area. Parts that have had a very long period between coating and molding may have to be freshened with a thin coat of primer.

Certain primers require a prebake prior to use. This can at times be achieved while the reinforcing is being heated in the mold.

6.2.4   Casting and Curing of Bonded Prepolymer

The polyurethane mix must flow over the part that has been primed and fully wet it. The presence of air pockets will prevent adhesion. It is most important that the primed part of the molding reaches the full cure temperature so that proper bonding will take place.

It is also very important to ensure that mold release does not come into contact with primer when the mold is being assembled. This will result in no bonding to all the contact areas. Care must be taken especially of overspray if the mold release is sprayed on.

When the bonded part is tested using the 90°Peel test (ASTM D429 method B), readings of over 53 kN/m (300 lb/in.) are usually obtained. When the bonding is good, the polyurethane will break (stock break) rather than actually peeling at the bond line. It is however recommended that the bond still be evaluated under appropriate field conditions. Without the use of the bonding primers, some adhesion is obtained but the strength is far too low for it to be considered an engineering-grade bond.

The strength of the bond will drop off as the temperature is raised. The strength decreases slowly until approximately 100 °C, after which it declines rapidly.

6.2.5   Post Casting

Post casting bonding of polyurethanes can be for two main reasons:

1. Repairs to existing parts, for example, filling in air pockets on surface of molding
2. Attaching polyurethane to another surface

Repairs

Filling in defects should be done as soon as possible. The surface must first be solvent cleaned to remove any traces of mold release and then slightly roughened with a barrel grinder. The area is then coated with some primer. A small dam is made around the part with a silicone sealant. A slightly catalyzed version of the same grade is used in the defect area. This process is best carried out between demolding and the final cure with some heat initially still in the part. Color matching is a major problem.

Bonding to Other Surfaces

Bonding cured polyurethanes requires that both surfaces be clean and free from contamination, especially traces of mold release. The surface of polyurethane should be roughened slightly and must be capable of making proper contact with the other piece. The polyurethane should be primed with the adhesive manufacturer's primer. The normal suppliers of polyurethane bonding agents can supply suitable adhesives and primers. Loctite also provides a range of bonding agents.

6.2.6   General

When the component is used in a chemical environment, the effect of the chemicals involved must be taken into account. Some cyanoacrylates have poor hydrolytic stability. If the bond line is not fully protected, the bond may break down over time.

6.3   Finishing

Most normal metal machining operations can be carried out on polyurethanes with some minor modifications to technique. With some experience polyurethanes harder than 90A can be machined relatively easily. Materials softer than 80A require more skill and experience.

The machining techniques used are like that employed with rubbers, for example, "knifing" and in extreme cases the polyurethane may have to be taken to subzero temperatures.

Using standard stock polyurethane sections, it is possible to machine prototypes using modified metal working techniques.

6.3.1 Differences from Metals

Before commencing any machining operations, it is best to consider how polyurethanes differ from metals and the general effect it has on the process.

Heating

The thermal conductivity of metals is much greater than that of all polyurethanes. The impact of this is that in the case of metals, the heat is readily removed by the coolant into the stock and machine. The lower thermal conductivity of the polyurethane means that the heat remains near the surface of the part. Without proper care the part can melt quite easily. Most polyurethanes start softening around 135 °C and melt to a gummy state by 180 °C.

Dimensional measurements taken when the material is hot will be much larger than when the material is at room temperature and 50% relative humidity. This means that parts that may be in specification if measured during machining may be too small when reinspected later. Experience is needed to know how much extra must be allowed so that on cooling the dimensions will be correct.

The use of water-based oil coolants and light oils will help to remove some of the heat generated at the cutting tool. Care must be taken to match the coolant to the next process in the production; for example, if the part is to be bonded or painted next, do not use any oil as it will have to be fully removed.

Memory and Modulus of Polyurethanes

When subjected to force, polyurethanes will distort. When compressed, the volume of the polyurethane remains, for practical purposes, the same. If the polyurethane is attached too firmly to the machine, it may be distorted. After machining when the part has been removed, the shape will change, giving a faulty part.

Polyurethane has a memory and when a force is removed, the material will return to its previous shape. A cutting tool will distort the polyurethane. After the tool has moved on, the material that has not been removed will return to its original shape. The swarf can foul the tool and generate more heat. The clearances on the tool must allow for the swarf to clear easily.

This memory effect can cause dimensions to change. Outside dimensions can increase while internal dimensions will decrease.

Safety Considerations

Standard safety precautions for the operation of the machine must be adhered to at all times and the guarding of the machine must be operational and used. Personal protective equipment must also be used. Eye protection is particularly important. The part must be correctly mounted on the machine with just sufficient force to prevent it from moving and coming loose during the operation. The tool and its holder must not interfere with the part when internal work is being done. This may cause the part to come loose under speed.

Do not inhale the smoke and/or dust generated during machining as it is dangerous to health. Smoke generated during any operation means that it is not being carried out correctly and the machining conditions must be changed immediately.

Softer grades of polyurethanes must be suitably supported to prevent undue distortions. Sharper tools with more clearance also need to be used.

6.3.2   Machining Conditions

Drilling

Recommended tools	High-speed twist bits with 90° or more points
	Thick materials angle 90 to 110°
	Thin materials 115 to 130°
	The drill should have slow spirals to allow easy removal of swarf
Size limitations	Nominal 9.5 mm diameter; smaller diameters should be clamped between plates to prevent distortion.
Grades of polyurethanes	All grades
Operation speeds	600 to 800 rpm
Work holding suggestions	Clamp or vice
Notes	(1.) It is hard to obtain close tolerances on softer grades.
	(2) Oversize drills may have to be used.
	(3) The drills must be kept very sharp.

Cutting, Shearing and Splitting

Recommended tools	Abrasive cutting machines with knife blades or shears up to 9 mm knife blade or water jet cutters
Size limitations	Sheets up to 13 mm
Grades of polyurethanes	All grades
Work holding suggestions	Friction-type clamps
Notes	Use bevel-edged knives to prevent uneven cuts.

Tapping

Recommended tools	Regular high-speed taps
Size limitations	None
Grades of polyurethanes	All grades
Operation speeds	600 to 800 rpm
Work holding suggestions	Clamp or vice
Notes	Softer grades very difficult.

Turning and Boring

Recommended tools	High-speed tool steel or carbide bits with a positive rake of 5 to 10°; must be razor sharp
Size limitations	Any diameter may be machined; length maximum 250 mm unsupported; softer grades will flex under cutting
Grades of polyurethanes	Increase rake with softer grades
Operation speeds	The speed should be high with a slow feed rate 70 to 89 Shore A cutting speed approx. 7 m/s 90 Shore A and harder cutting speed approx. 2 m/s
Work holding suggestions	Angle of cut to work 90 to 120°
Notes	Form tools can be used with the shore D grades. The chip material should be in a long, continuous strand. As the hardness increases, the surface finish improves. The tool should be just below the center line of the work.

Sawing

Recommended tools	Sharp 2 to 4 hook blades with raker set (teeth set to right and left)
Size limitations	200 mm diameter with cut 3 mm thick
Grades of polyurethanes	All grades
Operation speeds	Speed 1800 sfpm for shore D grades to 1200 sfpm for softer grades
Work holding suggestions	Hand or vices
Notes	Rotate round pieces to even out heat generated. Keep blade tension high to reduce friction on parts.

Milling

Recommended tools	Single-bladed flycutters or a two-fluted end mill cutter with 10° back rake and maximum clearance
Size limitations	Minimum of 9 mm thick
Grades of polyurethanes	Shore D grades the best (90 to 95 Shore A grades need extreme back clearance) Milling of material softer than 80A is not normally recommended
Operation speeds	Speed 900 to 1200 rpm with a 75 mm cutter. With a feed rate of up to 8 mm/sec
Work holding suggestions	Vice or double facing clamping tape
Notes	Fly cutting at approximately 75 mm diameter is the best.

Grinding

Recommended tools	The grinding wheels should be coarse in texture, and fine grained; carborundum (43 grit size) is a suitable medium
Size limitations	Wheel size to suit diameters of job
Grades of polyurethanes	All hardness; softer grades can deflect more
Operation speeds	Part approx. 150 rpm; some recommendations of reverse direction; grinding wheel 2250 to 3250 rpm
Work holding suggestions	Mount on lathe with grinder on tool post
Notes	Coolants such as water normally required in form of fine mist or brush. If no cooling — collect dust when grinding. Lower feed or use finer grit to obtain smoother finish.

6.3.3   Painting

Polyurethanes produced with aromatic isocyanates tend to yellow rather badly when exposed to the UV rays in sunlight. Internal pigmentation is not always successful in parts that are exposed to the sun. Polyurethanes pigmented with colored pigments (especially white) still change color drastically.

There are two main problems in obtaining good adhesion of paint to the surface of the part. These are:

1. Remaining mold release on the material
2. Plasticizers that migrate to the surface

The use of a nonsilicone mold release (such as a vegetable oil spray) in an aluminum mold may help. The mold release must be removed with a solvent such as isopropanol and water scrubbing.

The paint used must be flexible enough to withstand all repeated movements in the material. The most suited paint is a nonyellowing polyurethane paint that can normally be applied to the well-cleaned surface. If required, the surface can be primed with a special primer that is obtainable from the larger supplier's such as Loctite®.

6.4   Plasticized Polyurethanes

There are two methods to soften (plasticize) polyurethanes. The first is to use reactive plasticizers such as long-chain hydroxyl- or amine-based materials and the second is to use a compatible nonreactive plasticizer such as DIOP or Benzoflex 9-88.

Warning

The potential end use of the product must be considered when selecting the plasticizer that is used. Plasticizers migrate to the surface and may contaminate a product such as food or be a health hazard.

Systems can be designed to cure in the 25 to 35 °C temperature range. A catalyst is added to the part containing the curative. The amount of catalyst added must be adjusted to give the required gel and demold time.

The addition of plasticizer disrupts the hydrogen bonding in the system and hence the physical properties decrease. To overcome this, at low hardesses, some long-chained material can also be added. A simple system is shown in Figure 6.5 where some additional plasticizer such as DIOP is added to a 95 Shore A polyether prepolymer that has been cured with Ethacure 300. Similar systems can be developed with polyester prepolymers by the addition of an

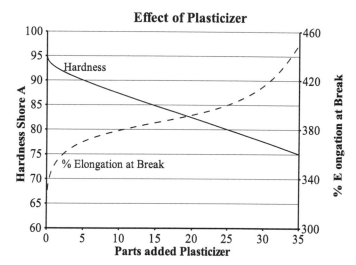

FIGURE 6.5
Effect of plasticizer on properties of polyurethane.

ester-based plasticizer. Polyesters are very much tougher than polyether so they make better softer materials. One commercial system uses a blend of a short-chain triol with either MOCA or Ethacure 300 and the addition of some plasticizer for the very softest material. The most used triol is Isonol® 93 which is now known as Conap® AH50.

When adding reactive plasticizers to polyurethane systems, the reactivity of the plasticizer must also be taken into account when calculating the amount to be added. The formula to calculate the equivalent weight of the blend is given in Appendix D.

Nonreactive plasticizers are held within the structure of the polyurethane. On prolonged heating they may slowly vaporize and the material will become harder.

The addition of phosphorous-based plasticizers will provide a degree of fire retardancy to the polyurethane. Fryol PCF and CEF have been used in polyurethane systems. Most phosphorus-based fire retardants are thin liquids that are compatible with polyurethanes. Due to the extremely varied nature of fire-resistant tests, the material made must be tested to the appropriate standard.

6.5   Epoxy Polyurethane Blends

Very hard polyurethanes can be brittle if not processed properly and need a degree of built-in thermoset properties to improve the compression set. Epoxy resins, when cured, are very hard. Their toughness can be improved by adding some polyurethane (of moderate hardness) to the epoxy resin during curing. The polyurethane affects the epoxy in the following manner:

- Increase in toughness
- Electrical properties still very good
- Chemical properties the same

One of the major reasons that the combination works is that both can be cured by the same aromatic amine curatives. The chemical structures also make them very compatible. It is considered by some that polyurethane acts as a plasticizer for the epoxy.

6.5.1   Curing Basics

The most commonly used polyurethane prepolymer is one of approximately 90 to 95 Shore A that is terminated with a TDI-based NCO group. The epoxy resins used are terminated with an epoxide group and have a molecular weight of 350 to 6000. The function of the amine curative is to

- Harden the epoxy by opening the epoxide rings
- Chain extend the polyurethane prepolymer

Heat is needed to activate the reactions and to reduce the viscosity to aid processing. The next major point is to select and adjust the curatives to obtain curing/chain extension at a similar rate. Diamine curatives such as MOCA and Ethacure 300 (DMDTA) can be used. Ethacure 300 is slightly on the slow side so a blend of Ethacure 300 and Ethacure 100 (DETDA) can be used. Ethacure 100 is very fast when used with polyurethanes on its own. The major difference in the usages of the amine curatives for the epoxy and polyurethane is that for the epoxy, both hydrogen atoms of the NH_2 group are used in the curing whereas for the polyurethane, only one hydrogen from each NH_2 group is used.

In the reaction between diamines and epoxies, all four hydrogen atoms in the amine group react with the epoxy groups; therefore the functionality is four. (With polyurethanes, each amine group reacts with an isocyanate group and hence the functionality is two). The basis of the calculations are that one epoxide group reacts with one hydrogen from the diamine.

Bond of one diamine to four epoxide groups

6.5.2 Processing

There are differences in the terminology used in the polyurethane and epoxy segments of the polymer industry:

- Epoxy Equivalent Weight (EEW)
- Epoxide number is the number of epoxide equivalents in one kilogram of resin
- The equivalent weight per active hydrogen atom is expressed as the amine hydrogen equivalent weight (AHEW)

$$\text{Equivalent wt.(g/mol)} = \frac{1000}{\text{Epoxide number}}$$

- This data is given for each batch

As with polyurethanes, the amount of hardener is calculated on 100 parts by weight (pbw). Epoxies are supplied with the epoxide value (EEW) of each batch.

The basic calculation is:

$$\frac{AHEW}{EEW} = \frac{X}{100}$$

or alternatively,

$$X = \frac{AHEW \times 100}{EEW}$$

For example, for an epoxy resin with epoxide value (EEW) of 183 if hardened with Ethacure 100:

$$\text{One would need parts by weight of E100} = \frac{(44.6 \times 100)}{183}$$

$$\text{for 100 pbw of resin} = 24.5$$

The required amount of curative for the polyurethane and the epoxy are calculated separately.

The amount of curative/hardener added is normally higher than the calculated value.

The major difference in the usages of the amine curatives for the epoxy and polyurethane is that for the epoxy, both hydrogen atoms of the NH_2 group are used in the curing, whereas for the polyurethane only one hydrogen from each NH_2 group is used.

This leads to different equivalent weights for the same curative. The main ones quoted are given in Table 6.1. The terminology used with epoxies differs from

TABLE 6.1

Curative Factors

	MOCA	Ethacure 100	Ethacure 300
Molecular weight	267	178.3	214
Equivalent wt. for:			
Polyurethane (Mol wt./2)	133.5	89.2	107
Epoxy (Mol wt./4)	66.8	44.6	53.5
(AHEW)			

that in the polyurethane industry. In a mixed system, the amounts of material required for each system are calculated separately. The following points should be considered:

- The reaction rate of E100 is the fastest.

- E300 needs lower temperatures than MOCA.

- The difference in the reaction rate between the polyurethane and epoxy needs to be taken into account.

The amount of curative/hardener added is normally higher than the calculated value. Up to 120% of the nominal amount for the epoxy is often used. Table 6.2 illustrates some of the properties of a start formula for a castable polyurethane epoxy blend. A 6.3% NCO TDI terminated prepolymer was reacted with a 200 epoxide equivalent difunctional epoxy. They were processed at 100 °C.

6.6 Cellular Polyurethane Elastomers

Standard castable polyurethanes can be converted into cellular polyurethanes by the addition of some blowing agent and surfactant that controls the bubble shape and size. These cellular products form a small part of the whole polyurethane foam family and the route is from castable polyurethane technology.

TABLE 6.2
Start Formulas

		Mix 1	Mix 2	Mix 2
Pu prepolymer		100	70	50
Epoxy resin		0	30	50
MOCA		19.1	28	33.5
Hardness	Shore A	95		
	Shore D	46	70	81
Tensile Strength	MPa	34.5	27.6	37.9
E@B	%	400	150	10
Brittle temperature	°C	<-70	<-50	-10

6.6.1 Rationale

These foams have a fine cellular structure and have improved vibration properties, shock absorbability and cut resistance (US Patent 3,929,026). The vibration damping and shock absorbability have led to its application as supplementary motor car springs and train-pad shock absorbers.

6.6.2 Basic Methods

The basic processing can be carried out by any of the polyurethane processing methods (one shot, quasiprepolymers or prepolymer route). The blowing (foaming) agent and silicone foam stabilizer are introduced into the formulation. The main point is that the foaming agent must be kept in solution until the mixture is in the mold. (This can be achieved by solubility in the mix, temperature or by pressure.) The heat of reaction of the system must be sufficient to vaporize the blowing agent and for bubbles to form.

The simplest blowing agent is water which will generate carbon dioxide gas on reaction with an isocyanate (This must be taken into account when calculating the amount of curative to use). Other agents that are used include methylene chloride or HFC-245fa blowing agent (1,1,1,3,3,-pentafluoropropane). Both require special handling. The size and shape of the cells are controlled by the use of silicone surfactants designed for the foam industry. (They are made by companies such as Dow Corning, Evonik or firms specializing in this type of product.)

Prepolymer Route

To produce foamed polyurethane, a liquid prepolymer based on PTMEG and TDI can be used. Typical start formulas are quoted in Table 6.3.

With adjustments to quantity, Ethacure 300 can be substituted.

Although polyurethane prepolymers are hygroscopic, it is difficult to add water directly so it is best added via triethylenediamine.

TABLE 6.3
Start Formulas for Elastomer Foams

	Formula 1	Formula 2
Adiprene® L100	100	
Adiprene® L167		100
Silicone surfactant	2.0	2.0
MOCA	9.7	19.0
Triethylenediamie (DABCO)	0.2	
Water	0.1 to 0.2	
Methylene chloride		8.0

Some ultrafine silica such as Cab-O-Sil® or Aerosil® (0 to 4 parts per 100 prepolymer) may be added to stiffen the foam as needed. These materials are very fine and need care in handling.

6.6.3 Quasiprepolymer and the One-Shot Route

MDI-based systems are more popular to use in the one-shot or quasiprepolymer route. To improve the compression set, some trifunctional polyols can be added (these are PPG-based polyols). The curatives, catalysts, blowing agents and foam stabilizers can be preblended. Points to consider:

- The material properties change. This may require the use of a harder grade of prepolymer.
- A balance must be obtained between keeping the temperatures down to prevent the blowing agent from flashing off and having the viscosity low enough to pour successfully.
- The curing reaction is exothermic (gives off heat), which is needed to carry out the blowing action.
- A triol may be needed to improve the compression set.
- The amount of blowing must only be sufficient to meet the technical demands of the polyurethane specification.
- Before methylene chloride is used, its strong solvent properties must be taken into account. It could have an effect on some seals, and it is a powerful paint stripper.

7

Polyurethane Processing Problems

7.1 Introduction

Faults encountered when hand-casting polyurethanes can often be a result of one or more problems. A complete review must be carried out when solving any irregularities. When defects appear in products that were previously good, it may be the interaction of several factors. The weather may have changed thus affecting the dew point. This may cause moisture to condense on the reinforcing. A second effect is the processing temperatures becoming too low.

It must be noted that the same cause may produce more than one different fault.

Key factors to reduce problems

1. All raw materials to be kept dry

2. All measuring devices to be operating properly

   (a) Scales

   (b) Temperature measuring and control devices

3. Nitrogen flow

4. Cleanliness of molds, reinforcings and workshop

5. Use of the right materials for the job

7.2 Observed Problems with Potential Causes

Bonding – bond failure to insert Lumps (bumps)when material hot.

Potential reasons:

Incorrect bonding preparation (see Section 7.5)

Bonding – poor bond to insert No strength in bond.

Potential reasons:

> Surface contaminated (see Section 7.5)

Cheesy Appearance Often MDI-based

Potential reasons:

> Low green strength (see Section 7.3.11)
> Insufficient cure (see Section 7.3.6)
> Off ratio (see Section 7.3.11)
> Poor mixing (see Section 7.3.9)
> Incorrect temperature (see Section 7.3.12)

Cracking of Part On demolding.

Potential reasons:

> Low green strength (see Section 7.3.11)
> Incorrect temperatures of mold and polyurethane (see Section 7.3.12)
> Wrong ratios (see Section 7.3.11)
> Poor mixing (see Section 7.3.9)
> Casting method (see Section 7.3.2)

Foaming In solid materials.

Potential reasons:

> Moisture in curative (see Section 7.3.3)
> Very moist prepolymer (see Section 7.3.3)

High Shrinkage Level of material lower than design in molds and vents.

Potential reasons:

> Incorrect temperatures of mold and polyurethane (see Section 7.3.12)
> High exotherm (see Section 7.3.4)
> Wrong ratios (see Section 7.3.11)
> System contaminated (see Section 7.3.3)

Physical Properties Durometer hardness, tensile strength, tear strength all low.

Potential reasons:

> Wrong curative ratios (see Section 7.3.11)
> Prepolymer degraded (see Section 7.3.8)
> Not fully cured (see Section 7.3.6)
> Poor mixing (see Section 7.3.9)

Incorrect temperatures (see Section 7.3.12)

Short Pot Life Faster increase in pot viscosity.

Potential reasons:

Curative contamination (see Section 7.3.3)
Incorrect temperature (see Section 7.3.12)

Snowflake or White skin Normally MDI systems.

Potential reasons:

Low green strength (see Section 7.3.7)
Insufficient cure (see Section 7.3.6)
Off-ratio (see Section 7.3.10)
Poor mixing (see Section 7.3.9)
Incorrect temperatures (see Section 7.3.12)

Striations Initial mix.

Potential reasons:

Poor mixing (see Section 7.3.9)

Striations Finished product.

Potential reasons:

Poor mixing (see Section 7.3.9)
Off-ratio (see Section 7.3.10)

Voids in Material Fine bubbles.

Potential reasons:

Moisture in system (see Section 7.3.3)

Voids in material Small bubbles randomly in part.

Potential reasons:

Pouring technique (see Section 7.3.2)
Dirty mold (see Section 7.3.4)

Voids in Material Large voids.

Potential reasons:

High exotherm (see Section 7.3.5)
Casting technique (see Section 7.3.2)
Dirty mold (see Section 7.3.4)
Mold design (see Section 7.4)

Wet Spots On surface of part.

Potential reasons:

Poor mixing (see Section 7.3.9)
Casting technique(see Section 7.3.2)

7.3   General Problem Solving

7.3.1   Avoid Forming Air Pockets

The mold must be designed so that the casting will fill from the bottom upward and not have a tendency to fold over itself. There must be sufficient air vents to allow the air to escape freely from the mold and not be trapped in any corners (for example, in flanges).

7.3.2   Casting Technique

Result
Air bubbles
Snowflake or white skin
Voids
Cracking
When pouring the polyurethane mix, care must be taken so that the polyurethane flows down to the base of the article and displaces the air as it fills from the bottom. If the polyurethane rolls over itself, it will trap air and form a void. The vent holes should be positioned such that the polyurethane can fill the whole mold. If required, additional holes should be inserted. The mold may have to be tilted to help with its complete filling.

If more than one mix is required to complete the pour, the second mix must be commenced while the first is still very tacky and not surface-cured, otherwise it will not fully blend in together and will form a crack or knit line.

A balance must be reached between various competing factors when casting polyurethanes. They are:

- The temperature of the mix and the mold must be suitable for the part.

- The viscosity of the material must be as low as practicable to allow easy filling and release of entrapped air.

- The polyurethane temperature must not be too high as to cause too fast a gelation.

7.3.3 Curative Contamination

Result

High shrinkage

Air bubbles

Short pour life

Unexpected foaming

Moisture in the curative or prepolymer manifests itself by very fine bubbles. Diols are particularly prone to moisture absorption. The moisture can be prevented from getting into the material by good housekeeping. The material must have a positive blanket of dry nitrogen over it at all times. If the material is properly degassed at an elevated temperature, these gas bubbles will be greatly reduced in number.

Care must be taken at all times to prevent cross-contamination of curatives by items such as catalysts, mold release or bonding agents.

7.3.4 Dirty Molds

Result

Air bubbles

Voids

All molds must be mold-released with a suitable mold-release agent. This agent must be applied very sparingly. If there is too much mold release, some of it may be wiped off during the pouring and cause faults in the polyurethane. Too little or no mold release can make the part virtually nonremovable from the mold. A mold, after several castings, will have sufficient buildup of mold release that one may not have to apply more for each subsequent casting. Only apply more mold release if one feels that the molding is becoming more difficult to remove.

7.3.5 High Exotherm

Result

High shrinkage

Snowflake or white skin

Voids

Cracking

Too high a mixing temperature will have two major effects, one being that the exotherm will be way too high and the second that the time available for casting will be too short. Too high an exotherm or processing temperature will cause high shrinkage. With MDI-based materials the surface will have a snowflake type effect or it may have a white skin on it. In extreme cases one may find a void or small crack internally in the part.

7.3.6   Insufficient Cure

Result

Low durometer readings

Poor tear

Cheesy appearance

Low tensile strength

The molded part can be removed from the mold as soon as it has sufficient strength not to break on demolding. The material should be post cured at the specified temperature, normally 100 °C for at least 16 hours. If this is not carried out, the physical properties will not be as high as they should. It must always be remembered that in the case of high-durometer TDI-based materials, although they are very hard soon after casting, they are very glass-like and will shatter readily. When they have had their full cure, they are both hard and tough.

7.3.7   Green Strength, Low

Result

Cheesy appearance

Cracking

MDI-based materials go through an initial cheesy state with very poor tear and tensile strengths. Once they have had their full cure, they are also very tough.

7.3.8   Low NCO Level

Result

Low durometer readings

Poor tear

Low tensile strength

The NCO level is normally specified by the prepolymer manufacturer but can slowly be reduced if the material has not been properly stabilized or if it has been stored for a long period at elevated temperatures. This will effectively give off ratio mixes, the consequence of which could be low durometer readings, poor tear strength, or low tensile strengths.

7.3.9   Mixing. Poor

Result

Low durometer readings

Wet spots

Poor tear

Cheesy appearance

Cracking

Striations

Low tensile strength

If the batch of material is incorrectly mixed there will be zones of material that will either have too much or too little curative. This will result in a number of potential problems, such as:

- Low hardness

- Mechanical properties being off specifications

- Internal stresses in the component

Very poor mixing will be evident by areas of uncured prepolymer appearing as wet sticky zones. Mixing containers should be cured off and checked for any wetness on the edges between the container and the cured polyurethane.

The mixing technique should be such that all the surfaces of the vessel and mixing utensils are well scraped off and that the curative is well mixed in. For smaller batches a spatula-type implement is normally used. Poor mixing is evident by the formation of striations in the mix. These are lines of different color or intensity that are evident both in the unpigmented and pigmented mixtures. The hand mixing should be such that the color or clarity of the mix is completely even.

For large batches an electric mixer may be used but in this case, care must be taken that air is not sucked down and whipped into the mix.

7.3.10   Off-Ratio

Result

Low durometer reading

Wet spots

Poor tear

Cheesy appearance

High shrinkage

Cracking

Striations

Low tensile strength

Off-ratio mixes can be a result of the NCO level in the prepolymer being too low.

If the prepolymer is stored at too high a temperature for an extended period, the isocyanate at the end of the chains reacts with itself to form multifunctional chains. Thus, when the curative is added, it will be in great excess as a result of the isocyanates being used up.

The fundamental calculations of the ratios of prepolymer, curative, and the rest of the system must be checked. The wrong EW or NCO levels may have been used. Alternatively the weights for a different curative could have been substituted. The correct functioning of the scales used must also be confirmed.

7.3.11   Ratios

The material can be off-ratio for a number of causes other than the low NCO mentioned earlier. These may be incorrect weighing or calculation of the amount to be added. Always make sure that the scales are checked regularly and that the calculations are verifiable and within the normal limits suggested by the raw material suppliers. There are a number of problems that are caused and they include low hardness readings, poor tear and tensile strengths. The pour life of the mixture will be quite different. If MDI materials are used they will have a more cheesy appearance with a higher shrinkage.

7.3.12   Temperature Incorrect

Result
High shrinkage
Voids
Short pour life
Cracking

Temperatures

High When the temperature of the mixture is too high the reactions will take place more rapidly. The heat given off by the reaction cannot escape due to the low thermal conductivity of the polyurethane. This can cause wrong chemistry. The initial viscosity of the mix will be lower.

The size and shape of the article will also affect the temperature required. A sphere may require a lower casting temperature to prevent overheating in the center.

Low If the temperatures are too low the viscosity of the prepolymer will make the mixing of the curative very much harder. The mixed prepolymer will also not flow easily into the mold and will tend to trap air against the mold surface. Another effect is that one can find a roll-type situation with the prepolymer not joining on itself.

Mold A missmatch of mold temperature against the prepolymer and the maximum exotherm can cause shrinkage. In a closed mold, internal pressures can cause damage.

7.4   Mold Design

The correct design of a mold aids in the speed of production and the lowering of the number of rejects. Figure 7.1 illustrates a typical multi-part mold to overcome some molding problems.

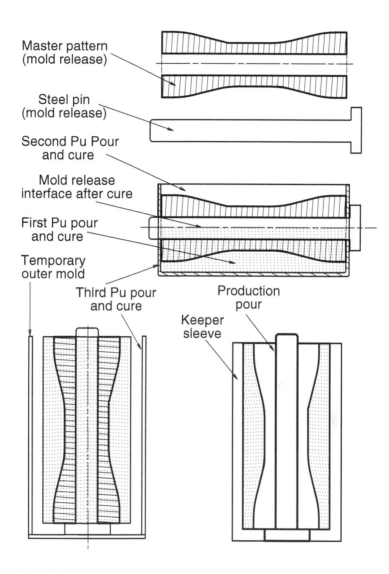

FIGURE 7.1
Typical mold design.

7.4.1   Use the Right Material and Finish for the Job

There is a wide range of materials that can be used to make a mold. The key factors are:

- The material must be dimensionally stable at the processing temperature.

- It must not give out gasses and moisture at processing temperatures (i.e. an unsealed coarse-grained timber). Materials such as plaster of Paris are not suited for hot cast molds.

- The finish that is required can be obtained.

The cost and finish of the mold must be economical in relation to the value of the work.

Pour Point Position

The pour point must not be on a good appearance side. This is normally the side the end user sees. In certain applications it may be the surface affecting the life of the part. In an open mold the pouring point should be positioned on a nonvisible surface or on one that will later be machined. The pour point will have a shiny, slightly dished surface. The same applies to any air vents.

Undue shrinkage in the mold is due to a mismatch in maximum exotherm temperature and the mold temperature.

7.4.2   Demolding Problems

Even a small undercut will prevent the easy removal of a part from a mold. A split mold can be used to overcome this problem.

Allow sufficient draft for easy removal. A very small draft, approximately 0.5°, will assist in the removal, otherwise a split mold will have to be used. Compressed air may be used down the side of a mold to help break the seal between the molding and the mold-release interface.

Sufficient mold release must be used to allow for easy removal of the part. Uneven application will have the potential to cause localized sticking. Too high an application is costly and can lead to pooling and possible knit lines in the polyurethane.

Water-based mold release must be fully dried and the mold temperature brought back up to temperature prior to pouring.

7.4.3   Shrinkage of System

Different casting and curing systems will give various shrinkage rates (from 0 to 1.5%). This must be taken into account when designing the mold.

7.5 Bond Failure

The type of failure can be established by examining where the bond has failed. Polyurethane will stick to a metal surface, but the strength is very much lower than a proper engineering-grade bond, where the strength will be greater than the tear strength of the polyurethane.

There is standard terminology representing descriptions for bond failures. These are:

RC	Failure at the polyurethane cement interface
CP	Failure at the cover cement primer interface
CM	Failure at the metal primer interface
R	Failure in the polyurethane

The polyurethane failure can be further broken down into a number of sub-groups. These classifications are based on normal rubber and that is why the **R** appears frequently. These include

SR Spotty polyurethane. It appears on the metal surface looking like splattered-on polyurethane. It is caused by either surface contamination with dust or that the sprayed-on adhesive has dried as it leaves the spray nozzles.

TR Thin polyurethane. There is an even but very thin residue of polyurethane on the metal surface. It is normally found with either butyl or polyurethane stocks that are very highly oil (or plasticizer) extended. The oil migrates to the surface and forms a layer that is part adhesive, part oil, and part polyurethane but is fairly weak.

HR Heavy polyurethane. A thick or heavy layer of polyurethane remains on the metal surface and indicates a very good bond.

SB Stock break. This is when the polyurethane appears to have folded back on itself and then broken off. This also shows very good bonding.

7.5.1 Bond Failures at the Metal-to-Primer Interface

Table 7.1 indicates the possible causes and solutions of metal-to-primer interface failures.

7.5.2 Bond Failures at the Polyurethane – Cement Interface

This failure mode is where a layer of primer is visible on the reinforcing but not on the polyurethane. The possible causes and solutions are given in Table 7.2.

TABLE 7.1
Metal-to-Primer Failures

Possible Causes	Remedies
Poor metal surface — oil or powdery residue evident	Better mechanical or chemical cleaning. Check degreasing operation.
Contamination of treated metal part before bonding agent application	Eliminate source of contamination. Cover or remove parts to clean area
Surface oxidized before application of bonding agent	Speed up coating process.
Manual handling of cleaned part	Automate or supply clean gloves for operators
Environmental destruction, i.e., salt spray, oils etc,	Check surface preparation techniques. Use robust metal primer,
Galvanic decay	Avoid dissimilar metal contacts upon installation of finished bonded parts.
Sacrificial metal activity	Avoid dissimilar metals in abrasion cleaning operation, i.e., steel grit with aluminum.
Bonding agent predried before reaching metal (spray application)	Reposition spray gun in relation to part.
	Check bonding agent viscosity. Add higher boiling point diluent.
Solvent trapped in bonding agent film	Increase time or temperature between application and bonding.
Destruction by solvents from protective paints or rust preventatives applied after bonding	Ketone types usually responsible.
Trapped air forced through bonding agent film	Improve mold ventilation.

TABLE 7.2

Polyurethane – Cement Failures

Possible Causes	Remedies
Precure of polyurethane	Pour polyurethane when the viscosity is low. Transfer molding — start transferring earlier.
Precure of bonding agent	Speed up transfer cycle. Lower molding temperature.
Leakage	Check mold for any sign of leaks. Ensure all split lines are well sealed.
Incorrect molding cycle	Check mold temperature.
	Check cure rate of polyurethane.
Low bonding agent film weight	Use heavier or multiple adhesive coats.
Migration from polyurethane	Substitute more compatible ingredients.
Contamination of bonding agent coated parts	Check unbonded specimens for oil or dust.
	Avoid overspray or excessive use of mold-release agents.
	Check for possible contamination from nearby equipment or agents.
Cement-polyurethane incompatibility	Use another bonding agent.

Part IV

Properties

8

Properties

8.1 Introduction

There are a number of major factors that influence the final properties of polyurethanes. The most important of these are:

- Type of backbone used

- Length of backbone

- Type of isocyanate

- Ratio of reactants

- Type and concentration of curative (chain extender)

- Final processing conditions

Careful choice of the polyurethane is needed to obtain a cost-effective article. Just as underspecifying an item can result in failure, overspecifying can result in the product no longer being viable.

8.1.1 Type of Backbone

Polyethers give good mechanical properties with excellent hydrolysis resistance. PTMEG (C4) ether is generally superior to those of the PPG (C3) ethers. The properties of polyurethanes made with PPG ethers have been brought closer to the PTMEG materials with the use of end capping with ethylene oxide (EO) and the low monal materials (Acclaim®).

Polyesters produce tough, oil-resistant polyurethanes with the major drawback being lower hydrolysis resistance compared to polyurethanes made using polyethers. The two newer groups of polyesters (polycaprolactone- and polycarbonate-based) both have better resistance to hydrolysis. Their toughness is very close to the basic polyester polyurethanes. Their disadvantage is cost.

8.1.2   Backbone Length

The length of the backbone will control the frequency of the hard segments that are present in the polyurethanes. The most obvious is the overall hardness of the material. The longer the backbone, the more flexible it will be. Short backbones with a degree of coordinate cross-linking (i.e., some short-length trifunctional groups) will produce material with high hardness and good compression set.

8.1.3   Type of Isocyanate

TDI-based polyurethanes produce the best properties when further chain extended with amine-based curatives. The overall properties can be increased if only the 100% 2,4-isomer is used. These materials are generally not suited for use with food. MDI-based polyurethanes have good overall properties and as they are predominantly cured with a diol, they can obtain FDA approval more readily.

In order to obtain an increase of 10 to 20 °C in usable temperature, isocyanates such as PPDI and CHDI need to be used.

Aromatic isocyanates, due to their structure, will yellow when exposed to light. The use of aliphatic isocyanates enables nonyellowing materials to be produced. The downside is the increase in cost.

8.1.4   Ratio of Reactants

Both the ratios in the prepolymer production and the curing ratios will affect the final properties of the polyurethane. In the initial prepolymer production, the properties vary according to the molar ratios of the prepolymer. This is illustrated by the graphs compiled in Saunders and Frisch [17].

The effect of varying the mixing ratios of the chain extender has a number of different influences on some of the more common properties. The changes depend on the effect at the molecular level.

The most commonly quoted property (hardness) remains relatively constant between 85 to 100% of the theoretical curative addition. This is due to the fact that the hard segments will provide the bulk of the stiffness to the product. Even when there is some hydrogen bonding, the hardness will stay approximately the same. Compression set needs a lower level of curative (85 to 95%) with some covalent cross-linking to develop the lowest set.

Properties such as abrasion resistance, resilience and heat buildup are normally best at a lower level of curative. These properties rely on unhindered flexibility of the backbone chain.

Tensile strength needs a level just below the nominal level (90 to 95%) as the strength relies on both the hard and soft segments without the effects of uneven chain lengths.

Purely physical properties such as tear strength, flex and elongation all require the curative index to be either at or slightly above the theoretical level. These properties need the strongest overall bonding in the matrix.

8.1.5 Type and Concentration of Curative

Within the same group of curatives some properties may be enhanced by using a certain curative. MOCA and Ethacure 300 both produce very good elastomers but may give certain properties that are superior to the others.

	MOCA	**Ethacure 300**
Nicked tear strength	Higher	Lower
Unnicked tear strength	Lower	Higher
Compression set	Lower (better)	Higher(worse)
Tensile strength	Lower	Higher
Production	Solid at ambient	Liquid at ambient
OH&S	Potential carcinogen	Not listed as such

8.1.6 Final Processing Conditions

All polyurethanes need the complete cure as specified to develop the properties fully. If the product is allowed to stand for a week at ambient temperature, the full properties will be developed.

A further heat treatment (sometimes referred to as annealing) or a longer, slightly hotter cure (18 hours at 130 °C), will also improve properties such as tensile, tear and overall toughness.

Very hard compounds (80 Shore D and above) need this extra heat treatment to fully develop the bonding network and to prevent cracking under load or impact. The products will form glasslike fragments unless the extra heat treatment is complete.

8.2 Physical Properties

There are a number of other physical properties that are very important to the optimum performance of polyurethane other than the normally quoted ones in trade literature. These include temperature, dynamic and hysteresis properties.

8.2.1 Temperature

Physical

The effect of temperature on polyurethanes has a number of zones [10]:

Below $-80\,^\circ$C	The material is a hard solid and in a glassy state.
-80 to $+20\,^\circ$C	The hard segments of the urethane begins to rotate and move.
20 to 130 $^\circ$C	The material is usable.
130 to 180 $^\circ$C	The polyurethane starts to soften severely.
Above 180 $^\circ$C	The polyurethane starts to break down irreversibly.

Within each of these zones the temperatures can be changed, depending on the exact nature of the backbone, isocyanate and curative.

The polyurethane becomes harder as the temperature decreases from 20 to 0 $^\circ$C. The environmental effects on the urethane decreases at below 0 $^\circ$C. At these temperatures the hardness, tensile and tear strengths and torsional stiffness increases. The largest increase is to the Young's modulus. The resilience of the polyurethane will decrease as the temperature is lowered.

The normal operating temperature of polyurethane is between ambient and 120 $^\circ$C (dry conditions). At these temperatures the properties are normally at their best. The envelope can be increased by the use of the most appropriate curative and backbone. For moist conditions the normal maximum temperature is 80 $^\circ$C using a PTMEG polyether, while it drops to 50 $^\circ$C for conventional polyesters. The newer esters have more hydrolysis resistance and can be used at slightly higher temperatures.

The dry and moist temperatures can be extended by using newer isocyanates such a PPDI.

The oxidative effects of air are greater on polyether-based materials compared to the polyester grades. This effect is due to the attack by oxygen on the ether bonds.

Dimensional

Polyurethanes have changes in their dimensions both during curing, and subsequent heating and cooling.

The shrinkage of hot-cast polyurethanes is a function of the temperature of the reaction. For a reaction temperature of 50 $^\circ$C, the linear shrinkage is approximately 1%. At 90 $^\circ$C it is close to 1.5%, and at 130 $^\circ$C it is 2.5%.

Cold-cast materials will have a very minimal shrinkage. This is useful where an exact-size copy must be made. Some post curing may be carried out using a stepwise process, for example, 4 hours at 50 $^\circ$C followed by 2 to 3 hours at 100 $^\circ$C.

It must be remembered that shrinkage takes place in all three planes. It is present but not normally detectable in thin cross-sections. In hollow items, the dimensions of the finished article will be smaller in outside diameter. The inside dimension will be larger. This must be taken into account when a mold is designed.

The coefficient of thermal expansion of fully cured polyurethane is between 11×10^{-5} and 8×10^{-5} in/in/ $^\circ$F. There is also an expansion of polyurethane at ambient temperature when the relative humidity increases. The major effect is between 70 and 100% relative humidity where it can swell up to 0.6%.

8.2.2 Resilience

The resiliency quoted for an elastomer is normally a single expression of the result of a specified test. The results quoted will vary according to the method used. To compare results, elastomers must be tested by the same method.

An explanation of resiliency is based on the conservation of energy. When an object strikes a polyurethane with a certain mass and velocity, the energy must remain the same. The mass of the object leaving the polyurethane is the same but its velocity is lower. The difference is retained in the polyurethane as heat energy.

$$M_1 \times V_1 = cf \times (M_1 \times V_2)$$

The factor "cf" is called the coefficient of restitution. A second explanation [20]

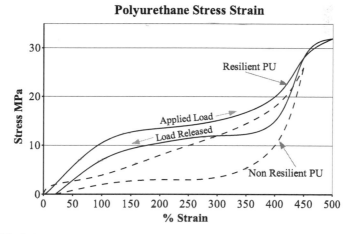

FIGURE 8.1
Polyurethane stress–strain curves.

is that when a stress (force) is applied, there is a small but measurable time lag before the material takes on the strain (change in shape). This is due to the need for the intermolecular forces to overcome the vibrational energy of the atoms. The warmer the polyurethane, the quicker the vibrational energy will be overcome and the polyurethane will become more resilient. If a stress-strain graph is produced, the applied stress and the recovery graphs are not the same. This is called hysteresis. Figure 8.1 illustrates the typical curves for a resilient and nonresilient polyurethane. The curves become more constant after three or four applications of stress. The molecules of all rubbers (elastomers), when allowed to rest undisturbed, align themselves in a dormant position. They become more active when some stress is applied. This can be illustrated if a ball that has not been used for some weeks is bounced. The first bounce will be much lower. After three to four bounces the rebound height becomes constant. The difference between the applied energy and the returned energy is stored in the part as heat.

The polyurethane can be considered to consist of two components, somewhat like the physics spring and dash pot model for viscous materials.

The elastic component (spring) stores and returns the energy. The second or viscous area (pot) converts the retained energy into heat. This is an important property in the design and selection of polyurethanes. The design of the polyurethane system can be adjusted to give varying amounts of resilience or absorption of energy. Figure 8.2 illustrates the spring and dash pot model.

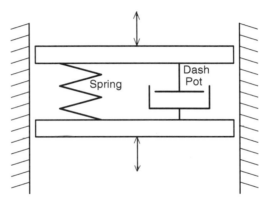

FIGURE 8.2
Spring and dash pot model for energy absorption.

The current method of determining the energy properties of polyurethane is the Dynamic Thermal Mechanical Analyzer (DTMA). This instrument applies a cyclic stress/strain to a sample of polyurethane in a tension, compression or twisting mode. The frequency of application can be adjusted. The sample is maintained in a temperature-controlled environment. The temperature is ramped up over the desired temperature range. The storage modulus of the polyurethane can be determined over the whole range of temperatures. Another important property closely related to the resilience, namely tan delta (tan δ), can also be obtained. Tan δ is defined in simplest terms as the viscous modulus divided by the elastic modulus.

In practice a material that is dead will store much of the applied energy and has a low resilience and a relatively high tan delta. A material that returns much of the energy has a high resilience and a low tan delta. These properties have an influence on the grades used for either shock absorbers, springs, bushes or wheels.

8.2.3   Thermal Conductivity

The heat generated when work is done to the polyurethane must be removed or the material will become overheated. The thermal conductivity of polyurethane is poor compared to metals. Unless the product is properly designed there may be failures in some polyurethane applications. The failure will manifest itself with the interior of the part melting and breakdown occurring from the inside of the part.

Thermal conductivity is expressed in several different internationally recognized ways. One method of expressing thermal conductivity (λ) is in terms of the heat flux under steady conditions per square meter for one meter of thickness of one degree Kelvin difference in temperature. Kelvin is a thermodynamic scale and is centigrade starting at absolute zero temperature.

$$\text{Thermal conductivity} = \frac{W}{m.K.}$$

Another method of expressing thermal conductivity is the quantity of heat passing through a unit area of the material, when the temperature gradient (when measured across unit thickness in the direction of heat flow) is unity [18].

The thermal conductivity of a polyurethane is in the order of 0.1 to 0.3 W/m.K. Other references give a value of approximately 1.7 to 3.5 x 10^{-4}cal.-cm/sec.cm^2 °C [9].

The low thermal conductivity of polyurethanes must be taken into account in the design of parts. The efficient dissipation of the heat must be allowed for when any part is subject to vibration, flexing or impact. The bonding of the polyurethane to a metal heat sink is often used to help dissipate the heat generated.

8.2.4   Stress-Strain Properties

The term "elastomer" is normally used to describe a material that will stretch when placed under load and will retract to approximately the original shape when the load is removed. The solid polyurethanes described so far have this elastic property. The elasticity differs from that of ordinary rubbers due to the nature of the bonds in the polyurethane matrix.

The chains that make a polyurethane "elastomer" have many polar groups with hydrogen bonding occurring throughout the chains. Natural rubber has many long entangled chains with some sulfur or peroxide cross-links. On applying a stress, the chains can straighten out and slide over each other. Reinforcing fillers such as carbon black are needed to reduce the elongation under load and produce a material with a high modulus. "Modulus" is the stress required to produce a certain strain. Expressed differently, it is the tensile stress (strength) at a specified elongation. Normally it is at 100, 200 or 300% elongation. Polyurethanes with hydrogen-bonded polar groups prevent the chains from sliding over each other and give a material with a higher modulus.

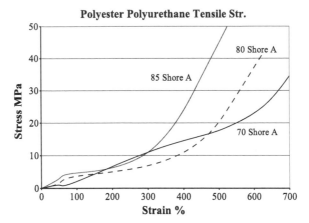

FIGURE 8.3

Tensile strength of polyester polyurethanes.

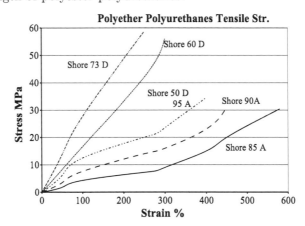

FIGURE 8.4

Tensile strength of polyether polyurethanes.

Polyurethanes "high modulus" means that reinforcing fillers are not needed to achieve the same properties.

It must always be remembered that polyurethane, like rubbers, has its tensile strength calculated using the initial cross-sectional area and not the area at break as with metals. Therefore the ultimate strength is much higher than conventionally quoted.

With metals, the modulus is stress divided by strain (Young's modulus) and is both a ratio and a constant. In the case of polyurethanes, the load defection curve is only linear over the first few percent. The Young's modulus is calculated in this area. As the curve passes through the origin, the modulus is the same in compression as in tension. Work has also shown that the Young's modulus is three times the shear modulus [20].

In practice, up to 90 percent of polyurethanes are used in compression, a few percent in torsion, and very little in tension. There is considerable data on the tensile stress against tensile strain (elongation) for polyurethanes. Most polyurethane specification sheets provide this data. Figures 8.3 and 8.4 show typical stress–strain curves for both polyester and polyether polyurethanes.

8.2.5 Compressive Strain

The compressive strain properties of urethanes show that polyurethanes have very good load-bearing properties. Softer materials (below shore hardness of 75 A) all have very similar response curves. The shape of these curves is influenced to a large degree by the ratio of the constrained polyurethane to the free area. This ratio is commonly called the shape factor. In calculating the shape factor, only the area of one loaded surface is taken. The total area that is free to deform (expand) is calculated. Figure 8.5 illustrates shape factors.

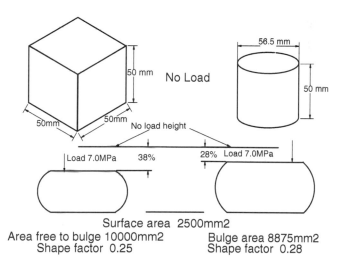

FIGURE 8.5
Shape factor determination.

The compression stress–strain graphs are normally determined using samples that are bonded to steel plates. If there is any lubrication, the results are completely different due to slippage that may take place.

Compared to conventional rubbers, polyurethanes retain their elastic properties at a Shore hardness of 85 A and above. At Shore hardness of 50 D, the elastic properties are still retained.

When the polyurethanes are in compressive stress as well as shear, the load-bearing properties are still retained.

8.2.6   Hardness

Hardness is a stiffness measurement. Stiffness is a stress/strain relationship. In load-carrying applications the stiffness is a bulk property, whereas measuring hardness is only a surface measurement. The hardness of an elastomer is a function of the number of closely packed hard segment clusters present per unit volume. The longer the soft segment chains the softer the material will be. The molecular shape of the curative will change the hardness to a lesser degree.

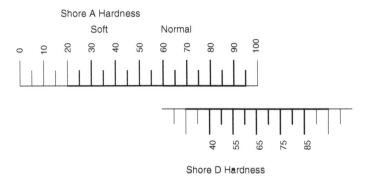

FIGURE 8.6
Shore A and D hardness scale overlays.

Hardness is measured by a specially shaped indentor that protrudes from a flat base. The other end is held against a spring. When held against a sample, the indentor is pressed into the case. The movement of the spring is measured either by analog dial or digital display. The harder the sample, the more the indentor moves into the case. In all scales, 0 is very soft and 100 very hard. The two scales used in polyurethanes are the A and the D durometer scales. Figure 8.6 shows the relationship between the Shore A and D scales. Two polyurethanes with completely different chemistry can have the same hardness. There are up to six completely different chemistries in common use (as shown in Table 8.1) that can give the same hardness.

In processing polyurethanes the hardness is not greatly affected by the mixing ratios. A material mixed at 80% of the theoretical curative level would have significantly different overall properties to one mixed at 100%. The hardness would still be at the same level. The hardness of polurethanes varies depending on the particular curative used (Figure 8.7).

To measure the hardness one needs to have a sample that is at least 5 mm thick and has a flat enough surface for the base of the durometer to rest on. For the most accurate measurements, a dead load instrument should be used.

TABLE 8.1

Basic Structure Combinations

Base	Curative	Comments
Ether polyol / TDI	Diamine cure	Both C3 and C4 subtypes
Ester polyol / TDI	Diamine cure	Traditional, caprolactone, poly-carbonate subtypes
Ether polyol / MDI	Hydroxyl cure	Both C3 and C4 subtypes
Ester polyol / MDI	Hydroxyl cure	Traditional, caprolactone, poly-carbonate subtypes
Ether polyol / PPDI	Diamine Cure	Mainly C4 subtype
Ether polyol / PPDI	Hydroxyl cure	Mainly C4 subtype

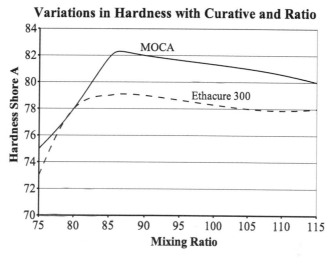

FIGURE 8.7
Effect of ratios and curative.

8.2.7 Tensile Strength

The tensile strength of polyurethanes is dependent on the resistance of the cross-linking bonds being either covalent or hydrogen bonds to prevent the chains from slipping completely and the material breaking. The harder the material, the more densely the cross-linking will be. This produces a material that is not only strong but also has a high modulus.

When operating below the ultimate tensile strength, the harder materials retain their elastic properties and do not behave like a thermoset. If a tensile test piece is allowed to recover, the sample will return close to its original size but there will be a degree of permanent set.

Polyurethanes under tension are very "notch" sensitive and when used as a spring, a nick will propagate and cause failure. Under long-term tension, polyurethanes will suffer from creep (strain relaxation) and set.

Fully post-cured polyurethanes will lose both tensile and modulus properties when heated in the short term (30 minutes) to temperatures above ambient and below their breakdown points. This effect is detectable both in classical testing and when using DTMA techniques.

These short-term effects are normally reversible at temperatures up to 120 °C with some changes to the hydrogen bonding possible. Prolonged heating will cause breakdown. This is shown in Figure 8.8. The elongation at break

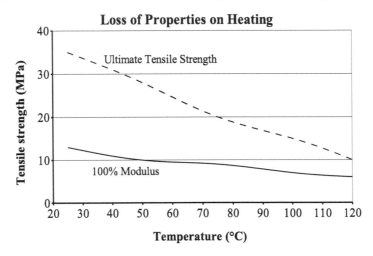

FIGURE 8.8
Loss of properties on heating polyurethanes.

increases as the temperature is raised. This is due to the secondary bonds being able to move more easily.

8.2.8   Tear Strength

The higher the modulus of the polyurethane, the higher the tear strength. The density of the cross-links appears to increase the tear strength. There are a number of recognized tests to determine the tear strength, namely nicked (notched) and unnotched. The nicked samples give a much lower tear strength result than the unnicked sample. Soft samples are very pronounced with the ratio approximately 10:1, while medium hardness (90 A) gives a ratio of 5:1. Hard materials maintain the same ratio as the medium hardness grades.

Polyester polyurethanes have higher tear strengths than their polyether counterparts. The nicked tear strength is much closer to that of polyethers.

As with the tensile strength, the tear strength decreases rapidly as the temperature increases. See Figure 8.8 in the tensile discussion.

Lubrication of the sample under tension while being torn reduces the tear strength greatly. This has a major effect in wear applications where fine nicked tears play an important role.

8.2.9 Coefficient of Friction

The coefficient of friction of polyurethanes has been found to be similar to that of rubbers. The coefficient of friction is the resistance to sliding or rolling of the surfaces of two bodies in contact with each other. It has been found that the softer the material, the higher the coefficient of friction. The values vary from 0.2 for the harder grades to approximately 3 for the softer grades. This is thought to be due to the higher actual area of contact between the elastomer and the second surface. A hard material under moderate load will not deform but will follow the surface profile of the second material. The coefficient of friction reaches a maximum at approximately 60 °C.

Laboratory tests must be taken only as a general guide. In practice, surface cleanliness and lubrication by dust, moisture and oil traces will greatly affect the actual friction properties.

The speed of friction will increase the coefficient slightly as it increases. The coefficient of friction will decrease over the course of time when the material is under load. It is thought that this may be due to the development of abrasive debris.

As the coefficient of friction can be reduced by the use of a lubricant, it is generally beneficial to use a suitable oil or grease when required. The addition of a modifier to the prepolymer itself must be done with care as the material will reduce the overall properties including aging, and may influence the bonding ability of the material. Additives that have been employed in this application include molybdenum disulfide, graphite and silicone oil. They must be used at the lowest level possible.

8.2.10 Compression Set

When polyurethane is subjected to a compressive force, it will deform. When the force is removed the material will recover some of its original shape. The amount of the deformation that does not recover is known as the permanent compression set. See Figure 8.9.

The normal method of evaluation is to measure the amount of permanent set with the constant strain of 25%. This means that the thickness is compressed by 25%. The ability of the material to recover is important in sealing applications where the compression forces may change due to temperature fluctuations.

Softer materials normally have better compression set results than the harder grades. This is due to the chains being able to move more readily over each other. Compression set values can also be improved by introducing some degree of permanent cross-linking at the prepolymer stage.

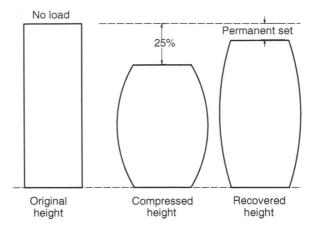

FIGURE 8.9
Determination of permanent compression set.

If polyurethane is placed under a constant stress, it will creep slowly over the course of time due to strain relaxation.

8.2.11   Permeability to Water

Polyurethane, like all other plastics, consists of chains that are always moving on the molecular scale. The polyurethane has molecular size voids in which it will allow small molecules to enter the material. The material breathes and allows moisture to enter.

This moisture has several effects. Initially it will react with any remnants of isocyanates and curatives. Moisture acts like a plasticizer, helping to give the final properties of the polyurethane. Under the influence of heat it will cause some breakdown of the material, especially in the polyesters.

When polyurethanes are used in the lining situation, this permeability of water can cause major problems in the form of blisters. The thin coating allows the moisture to reach the support layer. If the bonding is incomplete or the water attacks the bonding agent, some of the moisture can condense if the outer layer is cool enough. This has two effects. First, there is a greater driving force for more moisture to enter the polyurethane as the equilibrium has been disturbed. Second, as the metal support heats up, again the moisture is vaporized and there is a high internal pressure. The pressure will extend the plastic polyurethane layer and a blister will form. This action will also cause an enlargement of the disrupted bond area. See Chapter 9, Figure 9.7.

In all such applications the bonding surface must be correctly prepared, the correct bonding agents used and allowed to dry and, if needed, activated properly.

8.3   Environmental

Polyurethanes should have the right mechanical, temperature, chemical resistance and wear properties to be successful in any application. In real-world applications there is normally more than one influence that will affect the properties of the polyurethanes.

8.3.1   Thermal

Cold

As the temperature decreases, the degradation effect on the polyurethane chain also decreases to virtually nil. Physical properties change, but these changes are reversible. The major changes are:

- Increase:

    - Young's modulus
    - Hardness
    - Tensile strength
    - Torsional stiffness

- Decrease:

    - Resilience

The stiffness increases slowly from 20 °C to approximately −25 °C after which it increases rapidly.

At −30 to −40 °C there are molecular changes to parts of the polyurethane chain. These changes are dependent on the regularity of the molecule. The use of mixed polyesters will give lower results than the straight polyethylene adipate. The ester-based polyurethanes will become brittle around −60 to −80 °C. If polyester polyurethanes are held at low temperature, crystallization can occur. They are very much harder and much stiffer than normal. This state can be reversed by gentle heating or work such as slight flexing.

Polyether-based urethanes do not become brittle until approximately −87 °C. Du Pont polyurethanes (now Uniroyal) indicate an elongation at break of 200% at −73 °C [6].

Heat

If kept within the range of ambient to 110 °C there is a temporary fall-off in most properties, except the Young's modulus, as the bonds become weaker. For polyether-based polyurethane at 70 to 80 °C, the properties are only half of the original, while at 110 °C the value drops to about 20%. This gives a normal safe working temperature of 80 °C. Figure 8.10 illustrates the aging effect on different cure systems. This working range can be increased by the use of isocyanates such as PPDI in the polyurethane [5].

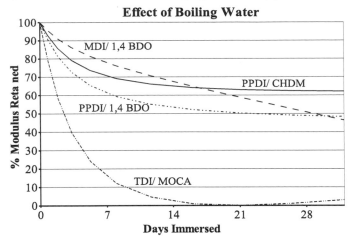

FIGURE 8.10
Effect of boiling water on different polyurethanes.

The second heat-related effect normally under dry conditions, starts at temperatures higher than 80 °C. Above 80 °C there is a gradual decrease in properties over time, and the rate increases with more elevated temperatures. Figure 8.11 shows the effect of dry heat aging on a typical polyether polyurethane.

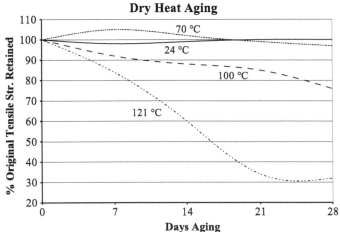

FIGURE 8.11
Dry heat aging of polyether polyurethanes.

Polyether polyurethanes are less stable at higher temperatures than the polyester-based materials. The ether linkage in the soft segment is attacked by the oxygen in the air while under the influence of heat. The order of resistance is:

$$\text{ester} > \text{urea} > \text{urethane} > \text{ether}$$

The mechanism is believed to be that a hydrogen atom next to the ether linkage is attacked. This radical reacts with oxygen from the air to form a peroxide radical, which in turn takes another hydrogen atom from the backbone of the chain to form a hydroperoxide. This hydroperoxide breaks down into two more radicals:

$$Ra^{\bullet} + H-R \longrightarrow Ra-H + R \xrightarrow{\cdot O\text{-}O\cdot} R-O-O^{\bullet}$$

$$R + R-O-OH$$

Hydroperoxide formation

When polyether polyurethanes are heated in an atmosphere of nitrogen, this thermal degradation does not take place and the material retains its properties.

8.3.2 Ozone Resistance

Ozone is found in the atmosphere especially around electrical appliances such as motors. Empirically it has been found that exposure to 3 ppm ozone under 25% strain gives a good indication of ozone resistance. At this level the material does not crack after 500 hours of exposure. If the level is increased to 100 ppm, some cracking is observed after 45 hours and the sample breaks after 460 hours.

8.3.3 Hydrolysis

Hydrolysis can be defined as the decomposition of a compound by reaction with water, the water taking part in the reaction. The effect is enhanced by the presence of either acids or alkalies. The chemistry of polyurethanes leads to the probability of hydrolytic attack. The mechanism is illustrated in the chemistry section.

Applications of polyurethane are often in moist air and water. Polyurethanes are often used in acid or alkaline solutions; even plain water is never actually neutral. Water is also present in the air and is absorbed into the polyurethane.

Early work by Athey [3] and Magnus [15] evaluated a range of polyurethanes for their resistance to moisture and dry and moist air environments.

Athey showed that polyether urethanes were five to ten times more resistant to hydrolysis than polyurethanes made from polyadipate urethanes. Further work showed that polycaprolactone polyurethanes were more resistant than polyadipate polyurethanes. Polyurethanes made with polycarbonate diols, have been found to be more hydrolytically resistant than the polycaprolactone polyurethanes. This leads to the series:

ether > polycarbonate > caprolactone > ester

Chemistry studies show that neighboring groups have an influence on the hydrolysis effect and diminishes as the ester groups become further apart.

Work showed that moist air (100% relative humidity) will cause more damage than moist conditions. This can be due to a combined effect of thermal oxidation and hydrolysis. In moist air the acid generated by the hydrolysis is not removed. The acid will catalyze the reaction further. In a liquid environment the acid generated will be washed away by the water solution.

Carbodiimides have been found to stabilize the polyester polyurethanes by blocking the carboxyl group formed when the chain has been broken. This helps prevent further autocatalytic degradation. The action is most pronounced when amines are used as the curative in caprolactone polyurethanes.

The loss in properties of polyethers at 70 °C (when in water) is evident with the life expectancy of twelve months. At temperatures above 50 °C the hydrolysis of polyadipate polyurethanes is such that their useful life is limited unless stabilization is used.

A ten-year long-term study was carried out in Panama on diamine cure polyether polyurethanes to examine their long-term resistance in wet and humid conditions. The ten years' exposure was found to have little effect on the samples that had been stored in the sea, jungle, soil, sun and in a hut. The samples exposed to the direct sun had the greatest loss of properties. Those in the sea had minimal loss in properties after the barnacles had been removed. [7]

8.4 Electrical

The basic electrical properties of polyurethanes are good but the polymer's tendency to absorb moisture can change the properties considerably. Polyurethanes are therefore not normally used as electrical insulators. They can be used in potting and encapsulating applications.

Any applications involving mains power must be evaluated and approved by the relevant local electrical authority before use in the application.

There are several important electrical properties. Insulation resistance is the resistance of polyurethane to the flow of electricity. The insulation resistance comprises of two main components. The "volume resistivity" is the

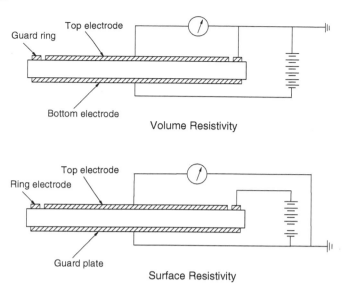

FIGURE 8.12

Volume and surface resistivity of polyurethanes.

resistance to flow through the bulk of the material and is a function of the composition of the polymer. The second component is the "surface resistivity," which is controlled by the condition of the surface of the polyurethane. See Figure 8.12. The cleanliness of the surface, purity (bleeding or surface breakdown), and the amount of moisture absorbed into the surface layer affect the surface resistivity. Moisture is slowly absorbed by the polyurethane and moves inward towards the center of the material.

The electrical resistivity of materials can vary from being an insulator through semiconductors to conductors of electricity. By careful adjustment to the chemistry and the use of additives, the resistivity can be controlled from being an insulator to a semiconductor. Figure 8.13 illustrates the change from a conductor to an insulator.

Arc resistance is the tracking of an electrical arc over the surface of a polymer. The arc will initially track through the air but depending on the composition of the polyurethane surface, decomposition of the polymer can take place and a more conducting track formed. The polyurethane will decompose to carbon that will readily carry the current. Figure 8.14 illustrates the diagrammatic method of how to determine the tracking resistance of polyurethanes.

Dielectric factors include the dielectric constant. This constant is a measure of the ability of a material to hold an electrical charge, as in a capacitor. The two other factors related to the dielectric constant are the "dissipation factor," which relates to the rate at which the charge is lost from the polyurethane, and the "power factor," which relates to the amount of heat generated in the storage of electricity.

Resistivity

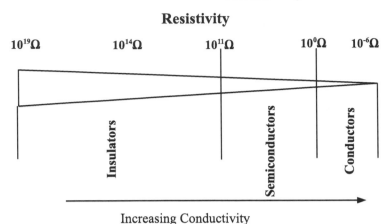

FIGURE 8.13
Resistivity spectrum.

The resistivity of polyurethanes changes both with temperature and relative humidity. Typical volume resistivity values at 50% relative humidity (RH), are 3.6×10^{12} ohm-cm at 23 °C, while at 100 °C it drops to 1.2×10^{11} ohm-cm. Another grade changes its resistivity from 3.5×10^9 ohm-cm at 20% RH down to 1.4×10^9) ohm-cm at 90% RH.

Polyurethanes find application where they dissipate electrostatic charges generated in rollers and items such as chute liners. There are two main routes to obtain these properties. The classical method is to add conductive graphite or fine metal to the polyurethane. The major disadvantage is the high loading of the material, which reduces the physical properties. The viscosity of the mix is also drastically increased and nonuniform mixing is often the result. The alternative is to use an anionic or cationic antistatic liquid. The agents are liquid surfactants such as quaternary ammonium salts, alkyl sulfonates, ethoxylates and ethanolamides. The liquid antistatic agents are hygroscopic and care must be taken to keep them dry. Some agents on the market are supplied in a water/alcohol mix. These are not suitable. The initial addition of up to 1% of the liquid has the greatest effect.

8.5   Radiation

Polyurethanes have been reported to be one of the most resistant polymers to gamma radiation. Early work by Harrington [12] comparing the effect to

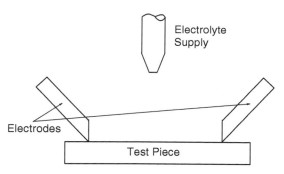

FIGURE 8.14
Electrical tracking of elastomers.

tensile strength and hardness, found that it is the total dosage rate that is the controlling factor. High dosages of 258 kC/kg (Roentgens) had some effect on tensile strength but the material was still serviceable. Initial work was done using thermoplastic polyurethane (Estane®). The exposure to the gamma radiation caused cross-linking in the material after initial chain scission.

The polyurethane was found to be more resistant to flex cracking than other polymers and retained its toughness and flexibility.

8.6 Chemical

The chemical resistance of polyurethanes allows them to be used extensively in industry. The individual chemistry of each grade must be used to optimize the use. For example, polyesters are tough and oil resistant. This makes them ideal for wiper blades on metal working machinery.

8.6.1 Inorganic Chemicals

Inorganic salts normally do not have adverse effects on polyurethane beyond that discussed under hydrolysis. If the pH of the solution is kept within the normal "neutral" range of pH 5.5 to 8.0, they have normally no catalytic effect on the degradation of the polyurethane. If the concentration is increased, the material will become either more acidic pH < 5.4 or more alkaline pH > 8.0. The acidity or alkalinity will promote the hydrolysis of the polyurethane. The temperature also plays a large part in the speed of the hydrolysis. Polyethers are also more resistant to hydrolysis than polyester-based polyurethanes.

In the following sets of data the grading is

A No/little effect
B Moderate effect
C Probably unsatisfactory
D Unsatisfactory

Inorganic Salts

Material	Grading	Comments
Ammonium nitrate	D	Acts more like nitric acid than a salt
Barium chloride	A	
Barium sulfate	A	
Barium sulfide	A	
Calcium chloride	A	
Calcium nitrate	A	
Copper chloride	A	
Copper cyanide	A	
Copper sulfates	A	
Magnesium chloride	A	
Potassium chloride	A	
Potassium nitrate	A	
Potassium sulfate	A	
Sodium chloride	A	
Sodium phosphate	A	
Sodium sulfate	A	
Salt water	A	Brine or sea water

Inorganic Acids

Strong acids like sulfuric and nitric acid have different forms of attack. Strong sulfuric acid will dehydrate the structure whereas nitric acid will oxidize it.

Material	Grading	Comments
Boric acid	A	Weak acid
Nitric acid, concentrated	D	
Nitric acid, dilute	C	
Oleum spirits	B	Sulfur trioxide in sulfuric acid
Phosphoric acid, 20%	A	
Phosphoric acid, 45%	A	Note upper operating temperature
Sulfuric acid, concentrated	D	
Sulfuric acid, dilute	B	

Alkali

Material	Grading	Comments
Ammonium hydroxide	A	
Calcium hydroxide	A	Very weak alkali - lime
Lye	A, B	Crude sodium hydroxide
Potassium hydroxide	A	Dilute solutions / Polyethers
Sodium hydroxide	A	Dilute solutions / Polyethers

Peroxides and Bleach

In dilute hydrogen peroxide (less than 10%) there is a slight attack, whereas with the strong peroxide (30%) the degradation is fast. This also applies to 12% bleach (sodium hypochlorite).

Inorganic Gasses

Material	Grading	Comments
Carbon dioxide	A	
Carbon monoxide	A	
Chlorine wet, dry or gas	D	Very aggressive gas will react with any moisture
Hydrogen gas	A	
Nitrogen	A	
Oxygen 95 °C	A	

8.6.2 Organic Chemicals

The term "organic" is used in the traditional chemical interpretation. Depending on the structure of the organic material, there will be varying effects on the polyurethane.

Alcohols, organic acids, ketones and esters can cause swelling and degradation at higher temperatures.

Hydroxyl

Material	Grading	Comments
Ethyl cellulose	B	Complex hydroxyl
Ethylene glycol	B	Complex hydroxyl
Glucose	A	Complex hydroxyl
Glycerin	A	Complex hydroxyl
Glycols	B	Complex hydroxyl
Ethyl alcohol (ethanol)	B	Simple hydroxyl
Methyl alcohol (methanol)	D	Simple hydroxyl
Whiskey, wines	A	Simple hydroxyl

Ketones

Material	Grading	Comments
Acetone	D	
Methyl ethyl ketone	D	50 °C

Esters

Esters and aliphatic hydrocarbons do not affect the polyurethane to a great degree.

Aromatics

Aromatic hydrocarbons need to be used with careful consideration. At ambient temperatures they can cause swelling and at higher temperatures there is a slow breakdown. It has been found that in uses up to 40 or 50 °C, polyurethane can be used in petroleum pump valve seats.

Material	Grading	Comments
Diesel oil	B	
Hydraulic oil	A	
Kerosene	B	Paraffinic
Petrol (with alcohol)	D	Alcohol makes it more aggressive
Petroleum > 121 °C	D	
Petroleum <121 °C	B	
Petroleum <71 °C	A	
ASTM Fuel 1 <70 °C	A	
ASTM Fuel A	A	Standard test fuels
ASTM Fuel B	B	Standard test fuels
ASTM Fuel C	C	Standard test fuels

Natural Oils and Fats

Natural oils and fats have little to no effect on polyurethanes. Use of food grades will often be required.

Material	Grading	Comments
Animal fats	A	Animal
Butter	A	Animal
Cod liver oil	A	Animal
Lard	A	Animal
Coconut oil	A	Vegetable
Corn oil	A	Vegetable
Cottonseed oil	A	Vegetable
Linseed oil	B	Vegetable
Olive oil	A	Vegetable
Peanut oil	A	Vegetable
Soybean oil	B	Vegetable
Stearic oil	A	Vegetable
Tung oil	B	Vegetable

Oils and Greases

Oils and greases find good application with polyurethanes at lower temperatures when the total wear, oxidation and degradation are taken into account.

Material	Grading	Comments
Asphalt	B	
Bunker oil	B	
Mineral oil	A	
Transmission fluid, type A	A	

Chlorinated Materials

Chlorinated solvents cause varying damage, from weak to relatively strong. Methylene chloride is very aggressive while carbon tetrachloride and trichloroethylene are relatively inert.

Material	Grading	Comments
Methylene chloride	D	This is a very aggressive solvent
Ethyl chloride	B	

There are numerous tables of chemical resistance available on the marketplace. There are a number of points to be considered prior to use:

- Concentration of material

- Whole composition of the material being used (e.g., a defoaming agent dissolved in diesel)

- Temperature of use

- Extreme conditions (e.g., dilute sodium hydroxide normally but strong caustic at 120 °for cleaning)

Before any final recommendation can be made, the entire application must be considered, both from a normal operating situation to the extreme conditions, including the changes to the material under operating conditions.

8.7   Wear

Wear is not a direct property of polyurethane but is the result of a complex system of materials being in contact, geometry of contact, operating conditions and environment [16].

Polyurethanes find successful use in applications where resistance to wear is required. The choice of the correct grade is vitally important as the resistance to wear is a complex problem involving knowledge of the wear mechanism, the wear environment and the properties of the grade of urethane.

There are two types of wear associated with polyurethanes namely:

1. Abrasive wear
2. Erosive wear

ASTM G40-13 "Standard Terminology Relating to Erosion and Wear" defines abrasion as "wear due to hard particles forced against and moving along a solid surface." Erosion is defined as the "progressive loss of original material from a solid surface due to mechanical interaction between that surface and a fluid, a multi-component fluid, or impinging liquid or solid particle."

Abrasive wear is a two-bodied wear that is found in a large number of applications where polyurethane is moving over a second object without lubrication. Wheels used on forklifts, trolleys or any other situation under load are typical examples of abrasive wear.

Erosive wear is a three-bodied wear that is found in pump and cyclone linings, impellers and screens. The polyurethane is attacked by a solid object that is being transported by a third medium such as process water. The temperature and chemical composition of the process liquid also plays an important part in the life of the component.

The process conditions, such as fluid composition and temperature, often have a far greater effect on the maximum usable life than the normal dry mechanical properties of the material. A typical change is from 120 °C maximum temperature for dry applications to 75 °C for wet applications.

8.7.1   Abrasive Wear

Abrasive wear is a complex combination of a number of factors including resilience, stiffness, thermal resistance, thermal stability, resistance to cutting

TABLE 8.2

Abrasive Wear Comparisons

Material	Abrasion mm^3.cm^{-2} (3.66 m)
Polyurethane 70 Shore A Ester	4.6
HD Polyethylene	6.7
UHMWPE	6.8
Nylon 66	7.5
PET	8.1
Polyurethane 80 Shore A Ester	9
Polypropylene	9.4
Polyurethane 80 Shore A Ether	10.1
Polycarbonate	11.8
Polyvinylidene fluoride	12.1
Polysulfone	12.5
Polyacetal	14.4

and tearing [18]. There are a number of laboratory tests, both international standards and commercial tests, for the evaluation of abrasive wear. The results from these tests normally only represent an indication of the actual wear that can be found in practice. The test equipment generally has a loaded sample against course abradant or in the case of a Taber abrader, a loaded abrasive wheel against a flat sample. (See Chapter 9.)

The wear rate can be considered to consist of three factors. The first is the properties of the material being worn; secondly, the angularity of the abrasive and lastly the nature and severity of the interaction of the abrasive; and the material being worn [11].

A comparison of the abrasive wear by Böhm et al.[4] found that the hard and more brittle plastics, such as polyacetals, polyamide-imide and polycarbonates, are less abrasion resistant than softer and tougher materials such as the polyethylene and polyurethanes. Table 8.2 details some comparative figures.

Within the polyurethane group of materials, the harder and tougher polyurethanes have the better abrasive wear resistance.

Trofimovich and Anisimov [19] considered polyurethanes to be a mixture of two insoluble materials, namely the hard and the soft phase. He found that with simple thermoplastic systems a relationship could be found with the density of the hard segment being a controlling factor. With the more complex cross-linked materials, the relationship was harder to establish. In filled compounds, the filler can protrude above the surface and change the wear conditions. The abrasive wear will reach a minimum depending on the concentration of the hard segment and the type of backbone.

Examination using a scanning electron microscope (SEM) of the surface of abraded polyurethane shows cutting and gouging marks on the surface.

The surface can deteriorate further, and fatigue cracks start to appear. If the conditions are extreme, the surface can soften drastically and molten/decomposed polyurethane can be seen under the SEM.

The heat generated by the abrasive wear is not as readily dissipated as with the erosive wear.

8.7.2   Erosive Wear

Erosive wear can be considered an interfacial wear where the energy evolved in the wear is dissipated. The third component in the wear process removes some of the frictional heat generated when the wear particle strikes the surface of the sample.

Erosion studies have shown that the softness and resilience of elastomers handle wear better than hard metals in a number of cases [8]. In dry applications the heat dissipation is poor and can result in rapid degradation of the elastomer. However, in slurry applications the lubrication of the water changes the friction resistance greatly.

Mode of Wear

Arnold and Hutching [2] found that the normal erosive wear of elastomers is characterized by fine cracks being formed under impact. These grow incrementally under cyclic impact loading. When these cracks intersect, small particles are removed. Cracks will form and grow when the rate of elastic energy increases due to impact and exceeds the surface energy associated with crack formation. As cracks grow, elastic energy is released.

Microcutting of Polyurethane

A mode of wear is microcutting where a sharp edge will cut through the surface. The general mechanism is shown in Figure 8.15.

Polyurethane under tension will wear at a greater rate than one that is under slight compression. This is similar to other polyurethane properties such as cut and ozone resistance.

An elastomer with a low modulus will often have far better erosive wear than material with a higher modulus. An abrasive wear test (such as the DIN abrader) will show a poor result for a soft elastomer. This is also shown in field applications such as tire wear. When the application is changed from abrasive to erosive wear, the softer elastomer will wear very well. The reason for this is that the low modulus of the soft elastomer allows the stresses from each impact to be dissipated more readily than for hard polyurethanes. The soft material will stretch further and then snap back before any damage is done. Any microcracks formed will have a slower growth rate and hence less erosion will occur.

Resilient elastomers will give a pattern of ridges approximately 15 to 30

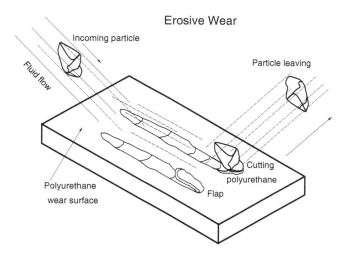

FIGURE 8.15
Erosive wear mechanisms.

micrometers apart across the elastomer at right angles to the direction of wear [13]. Attached to the ridges will be globules of worn polyurethane. Low resilient elastomers have higher wear rates. The surface is rough and does not show the same ridges.

Studies of the properties of polyurethanes, both physical and physico-chemical, showed only that resilience had some direct relationship to wear. A relationship of $(1 - \text{resilience})^{1.4}$ where resilience is expressed as a fraction has been shown. No other relationship has been found for other properties for erosive wear between 30 °and 90 °s [13].

There are a number of external factors that affect the erosive wear of the elastomers. These include:

- Angle of impact

- Particle velocity

- Particle size

- Particle shape

- Presence of lubricant

Angle of Impact

The angle of impact of the particles on the polyurethane has a major effect on the erosive wear. Both Hutching and Deuchar [13] and Li and Hutching [14] found that the wear of polyurethane was ten times greater at 30°s than at 90°s. The wear mechanism did not change with different mechanical properties.

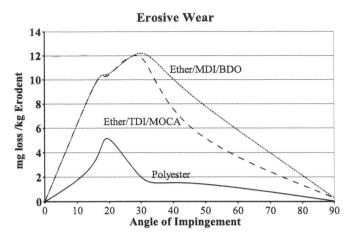

FIGURE 8.16
Effect of impingement angle on erosive wear of polyurethanes.

Figure 8.16 shows the erosive wear variations between impingement angles and prepolymer type.

At 90°s impact angle the mode of wear is one of many direct cuts and not the rolling of the surface. The number of interparticle collisions also increases greatly.

Velocity of Particles

At high particle velocity (above 50 m/s) the wear rate increases more than what would be expected from straight kinetic modeling[2]. The reason is believed to stem from higher formation and propagation of microcracks.

The rate and amount (flux rate) of material impinging on the elastomer has an effect on the wear rate of the elastomer [1]. With flux rates of between 500 and 5000 kg m^{-2} sec^{-1}, the erosive wear decreased with flux rate at low velocities and low angles of incidence. At higher flux rates (1000 to 10000 kg m^{-2} sec^{-1}), the wear rate decreased when the velocity was lower and the angle of impact was more normal to the surface.

At the high flux rate there will be a large number of particle-to-particle collisions that will decrease the wear rate. Water and air will be forced into the surface of the polyurethane. Hydroperoxides can be formed under these conditions, resulting in breakdown of the polyurethane in a similar manner to that described for hydrolysis.

Particle Size

The load applied to the surface of the polyurethane by the impinging particle is considered an important factor in the wear process [8]. The force is a product of the mass and velocity. The mass of the particle increases greatly as its size becomes larger (the volume increases by the cube of its dimensions). When a certain force is reached, rolling wear will commence. Depending on the force, the elastomer will either deform or, if the force is large enough, a particle of polyurethane will be removed in a scallop-like piece or a curled-up leaf.

Particle Shape

Particles can be any shape, from round to knife- or dagger-shaped. The shape can also change during the wear process. The way particles break on impact is very important. The edge of the particle can vary from being blunt to razor sharp. Sharp-edged and flaky particles are considered the most abrasive[8]. The sharp edges often have a slashing effect.

Presence of Lubricant

The largest effect of water or any other liquid is that it acts as a lubricant and reduces the erosive rate by reducing the frictional forces. This effect is greatest at angles between 30°s and 90°s. If water instead of air is used, the wear rate is reduced by 50% [2]. To a lesser degree the lubricant will enhance any cutting that may take place. The formation of hydroperoxides is another effect caused by water on the polyurethane.

8.7.3 Summary

The wear of polyurethanes is a complex combination of a large number of influences, both from the polyurethane and the material that is doing the wearing. Over the course of the life of the polyurethane, the properties can change and the wear rate altered. Flow patterns can also change, affecting the wear to a large degree.

References

[1] J. C. Arnold and I. M. Hutchings. Electron beam damage in scanning electron microscopy of worn elastomer surfaces. *Wear*, 128:339–342, 1988.

[2] J. C. Arnold and I. M. Hutchings. The mechanisms of erosion of unfilled elastomers by solid particle impact. *Wear*, 138(1-2):33–46, June 1990.

[3] R. J. Athey of DuPont. Adiprene water resitance of liquid urethane vulcanizates. *Trade literature brochure*, pages 1–8, 1965.

[4] H. Bohm and S. Betz. The wear resistance of polymers. *Tribology International*, 23(6):399–406, 1990.

[5] J. Chin and A. Sing. p-Phenylene diisocyanate for high performance castable polyurethanes. In *Rubber Division, American Chemical Society*, pages 1–21, Nashville Tennessee, 1992.

[6] Uniroyal Corporation. Adiprene L167. *Trade literature brochure*, pages 1–35, 1993.

[7] D. W. Cumberland. Panama weathering study by DuPont Corporation. *Trade Literature*, pages 1–22, 1985.

[8] H. J. Ephithite. Rubber linings — The soft options against abrasion. *Bulk Solids Handling*, 5(5):1041–1047, 1985.

[9] Gallagher Corporation. Cast and molded componets of polyurethane elastomers - design and application guide. *Trade literature brochure*, page 35, 1994.

[10] J. T. Garrett, R. Xu, J. Cho, and J. Runt. Phase separation of diamine chain-extended poly(urethane) copolymers: FTIR spectroscopy and phase transitions. *Polymer*, 44(9):2711–2719, April 2003.

[11] J. Gates. Abrasive wear — theory and practice. In *IMEA/ WTIA Joint Seminar on Abrasive Wear*, Brisbane, 2001. IMEA/ WTIA.

[12] R. Harrington. Radiation effects on polymers. *Rubber World*, 88(3):4–5, 1960.

[13] I. M. Hutchings and D. W. T. Deuchar. Erosion of filled elastomers by solid particle impact. *Journal of Materials Science*, 22:4071–4076, 1981.

[14] J. Li and I. M. Hutchings. Resitance of cast polyurethane elastomers to solid particle erosion. *Wear*, 135:293–303, 1990.

[15] G. Magnus, R. A. Dunleavy, and F. E. Critchfield. Stability of urethan elastomers in water, dry air and moist air environments. *Rubber Chemistry and Technology*, 39(4):13 – 28, 1966.

[16] J. I. Mardel and K. R. Chynoweth. The wear of polyurethane elastomers. *Materials Forum*, 19:117–128., 1995.

[17] J. H. Saunders and K. C. Frisch. *Polyurethanes Chemisty and Technology Part One*. Interscience Publishers, New York, 1962.

[18] L. P. Smith. *The Language of Rubber*. Butterworth Heinemann Ltd., Oxford, 1st edition, 1993.

[19] A. N. Trofimovich and V. N. Anisimov. Role of structure factor in evaluating polyurethane wear resistance. *Trenie I Iznos*, 8(3):493–499, 1987.

[20] P. Wright and A.P.C. Cumming. *Solid Polyurethane Elastomers*. Maclaren and Sons, London, England, 1st edition, 1969.

9

Applications

9.1 Introduction

Polyurethanes find application in a wide range of industries in competition with other plastics, ceramics and metals. Cast polyurethanes are tough, abrasion resistant, have load bearing and are chemically resistant. That makes them first-choice materials in certain situations. The ability to be easily prototyped and have short production runs is a major positive factor for polyurethanes.

The correct choice of the type and grade of polyurethane is important in producing a successful product and a happy customer. Some of the important criteria are:

- What is the main purpose of the application?

- Are the conditions static or dynamic?

- In what environment will the part work in?

  - Heat
  - Cold
  - Chemical
  - Food/drugs
  - Electrical, radiation, etc.

- What are the expected extremes in working conditions?

- Will there be any abrasive or erosive wear?

- Is there a potential internal heat buildup?

- Is there any history of previous use of polyurethanes in similar situations?

When selecting polyurethanes for any application, a full understanding of the working conditions should be obtained to prevent over-specifying the grade and type. These may include the interaction of temperature, hydrolysis and wear in a dynamic situation. The limitations of polyurethanes must also be taken into account and some redesign carried out if needed.

9.2 Major Type and Grade Selection

A number of traditional factors are often specified for polyurethanes, such as hardness and tensile strength, that do not control the final performance properties.

Hardness

The hardness of polyurethanes can be obtained either by the density of the hard segments, the choice of curative or the addition of plasticizer. The importance of hardness is to control the rigidity of the part (if not reinforced) or its softness to conform to the shape of some other object it passes over.

For very low hardness (<60 A) applications such as rolls, TDI esters are the most suited whereas ether-based materials are least suited.

Hard materials do have better abrasive resistance than soft materials.

Tear Strength

Tear strength is important where there is any tension or the potential for "nicking" of the surface. The structure and internal bonding in the polyurethane give the tear properties. The high toughness of the esters makes them most suited while materials made from the low-cost ethers have the worst resistance.

Temperature

Low-temperature applications are controlled by the stiffening of the polyurethane as the temperature goes from 20 °C down to −20 °C. In these applications MDI ethers are the most suited while ester-based materials become very stiff. At very low temperature, esters have a lower brittle point than ethers but they must be kept from crystallizing.

In high-temperature applications the use of the newer isocyanates such as PPDI and CHDI provide a higher maximum temperature range of up to 110 to 120 °C dry but they do come at additional cost. For materials in the slightly lower temperature range, TDI-based materials give better heat resistance than MDI-based materials.

With sustained heat, aging esters will maintain their properties longer than with ethers. The older style PPG-based materials have the least heat resistance.

Heat Buildup

Heat buildup is closely related to resilience and a low tan δ value. The design of the part is vital to allow heat that is built up to dissipate successfully. Ether-based materials have a lower heat buildup than esters.

Hydrolysis Resistance

In any choice of material involving hydrolysis resistance, the time – temperature relationship with regard to hydrolysis must be known and appreciated. The newer types of esters offer better resistance than the traditional grades. This means that the newer material can operate at higher temperatures than the previous maximum of 50 °C.

MDI-ether-based materials are classified as the most hydrolysis-resistant polyurethanes.

Compression Set

Compression set is controlled by the cross-linking in the material. This may be carried out during the design of the prepolymers or by using some short chain cross-linking agents. TDI-based polyurethanes have better compression sets that the MDI-based materials.

Tensile Strength

Polyurethanes are occasionally used in the tensile mode. In these applications the strength and toughness of ester-based materials are superior to those made from ethers. The ultimate elongation of all polyurethanes far exceeds any application so the choice is not of any major consideration.

Oil Resistance

The oil and grease resistance of the ester group of polyurethanes, together with their toughness, makes them most suited to these applications.

Wear Resistance

For sliding abrasive resistance the tough polyesters are the best suited. This must also be balanced against the hardness of the material. The older style PPG materials do not have the physical properties to provide outstanding resistance to abrasive wear.

When there is impinging erosive wear, MDI-based polyethers are normally classified as the most suited due to the fact that hydrolysis resistance is also taken into account. Certain more sophisticated esters have better erosive wear than the high-cost ethers (PTMEG) under hydrolysis conditions. The temperature limitation however is still just below that of the ether-based materials. The angle of impingement must also be taken into account.

Food and Drug Approval

MDI-diol-cured polyurethanes generally can receive FDA approval for use provided no mercury catalysts are used. Polyurethanes made from most amine-cured TDI material will not meet the requirements for FDA approval. Trimethyleneglycol di-p-aminobenzoate (Versalink™ 740M) of the diamine curatives has however been approved. If additives (e.g., catalysts and plasticizers) are used with the polyurethanes, these must also be considered in the approval process. Appropriate local and state approvals must be obtained prior to use where food or drugs come in contact with the polyurethane.

Flexibility in Adjustments to Formula

Polyurethanes based on MDI have the greatest flexibility regarding choice of curatives and blends of curatives that can be used to obtained the desired properties. In castable elastomers the hygroscopic nature of the curatives must be taken into account and proper handling and storage used; otherwise there may be a very high reject rate.

Catalysts are also more readily used with the MDI materials. Organic acids such as adipic or oleic acid work better with TDI materials than with MDI systems.

TDI-based materials are normally cured with a range of amine-based curatives. The currently most popular are Ethacure 300 followed by MOCA. MOCA is still popular despite the carcinogenic concerns associated with it.

For soft materials, extension with either a reactive or nonreactive plasticizer is best with an ester-based material. The toughness of the ester helps keep the properties usable at lower hardness.

Cost

The unit cost of polyurethanes varies greatly from the lowest raw material cost of the PPG/TDI-based systems to a heat-resistant, optically clear aliphatic grade. The actual lowest volume cost is the cost of a system that will fully meet the needs of a specified item. The use of a grade that is too good for the application wastes money. Conversely, a grade that is not suited also costs money.

9.3 Polyurethane's Role in the Materials Field

Polyurethanes have several advantages over competing materials in the materials field. The major items of competition are metal, other plastics and rubbers. Ceramics offer some competition to urethanes. Each of the above groups of materials, including polyurethanes, requires its own design adjustments for successful use in any application.

9.3.1   Comparison to Metals

The very much lower density of polyurethane (1.0 to 1.2g/cm^3) compared to the lightest of standard metals, namely aluminum ($\sim$2.7 g/cm^3), gives it great weight advantages. The difference is magnified even more when compared to steel. Parts can be made so much lighter that they can be handled with ease.

The fabrication of polyurethanes into complex shapes is much easier than with metals. Large casting of up to 500 to 1000 kg can be made from relatively simple molds, using buckets and minimal labor. The energy input into the production of a cast polyurethane part is low compared to the melting of a metal alloy. The actual molding costs are also lower.

Polyurethanes provide a large package of chemical resistance to metals within the operating temperature of the polymer. By careful selection, alloys can be used to provide the desired chemical resistance but usually at a cost disadvantage.

Choosing polyurethanes with the correct resiliency can be made to either be resilient or energy absorbing. The operations do not produce the noise levels that metals produce when they strike each other. The design of the polymer parts must be such that the heat buildup, due to the lower heat conductivity, is removed. This is often achieved using a thinner cross-section and bonding the material onto metals.

For smaller particle sizes (∼1000 microns), polyurethanes provide a superior erosive wear resistance to metals at most normal velocities of up to 20 m/s. Outside these limits the materials need to be evaluated in a manner that is as close as possible to real conditions.

Polyurethanes can flex and deform to assist in any movement of parts. The elastic nature means that the polymer will return close to its original shape after any minor deformation. This flexibility can be a disadvantage in certain situations and the positive attributes (e.g., better wear), nullified. In such situations the casting of an outer surface of polyurethane over a metal reinforcing is advantageous.

The chemical structure of the polyurethanes make them relatively non-conductive and with suitable modification semiconductive, allowing for the controlled discharge of static. Polyurethanes will not produce a spark when struck by another object.

9.3.2 Advantages over Plastics

Within the polymer group there may be certain plastics that have specific individual properties that are superior to polyurethanes. In the overall situation, the total requirements of the application must be considered.

The ability to produce polyurethane parts with a large cross-section (>30 mm) easily is a major advantage over other polymers. A conventional polymer such as polyethylene requires heat under pressure to form a thick cross-section with a long, slow cooling/annealing cycle to prevent stress and voids in the part. The production of other than simple-shaped articles is very difficult. After this molding process, post machining also needs to be carried out.

Tooling costs for the production of all polyurethane parts are very much lower than those for making plastic parts by either compression or injection molding. The molds do not have to withstand the pressures involved when compression and injection molding are carried out. These lower molding costs can be used to prepare prototypes of a new concept for general evaluation prior to the expense of an injection molding die.

The wear resistance is better than most plastics, except for UHMWPE in sliding wear applications such as chutes. Polyurethanes can be made in thick-sectioned intricate shapes. This makes them a very good choice in wear-resistant applications.

The thermoset component in polyurethanes gives them a better compression set than most thermoplastic polymers. They also have better cold flow properties. Polyurethanes are tough and more resilient than a large number of other plastics.

The radiation resistance of polyurethanes is better than that of other polymer materials and makes them a good candidate in applications where there is gamma radiation.

Most polyurethanes need UV additives to stop yellowing, except for the specialized nonyellowing clear grades. These grades are far more expensive than the standard materials.

9.3.3   Advantages over Rubber

Rubber can be formulated to meet most of the properties of polyurethanes. This is a complex capital- and labor-intensive procedure. In most applications more than one property is needed, and polyurethanes provide many of these at one time.

The capital and process control costs to prepare rubbers must be considered against the cost to process the polyurethanes. The straight raw material cost of the standard rubbers (natural, SBR and chloroprene) will be less than the polyurethane's cost but the overall processing cost of the polyurethane will be lower.

In all these comparisons the overall properties must be taken into account. Even though fluoroelastomers have excellent compression set properties, the other poor physical properties and cost must also be considered.

The load-bearing properties of polyurethanes are superior to those of conventional rubbers. The properties of polyurethanes at harder hardnesses are superior to those of the rubbers. To reach the 60 D to 70 D range, the conventional rubbers must be very highly loaded or in the form of an ebonite (25% sulfur). Polyurethanes can be produced in this high range while maintaining the elastic properties that are also present in the lower hardness materials.

Rubbers rely on fillers (both reinforcing and nonreinforcing) to obtain their properties. The curing system also produces a dirty-colored material. To color a rubber is difficult and only a few basic colors are used. To obtain a transparent rubber, special latex or synthetic cis-polyisoprene must be used, and the use of a peroxide cure is normal. Polyurethanes can be colored any color but the yellowing of aromatic systems must be taken into account. Aliphatic systems can give transparent nonyellowing systems.

Rubbers that have the same or better properties than polyurethanes are detailed below:

Property	Rubber group
Resilience	Natural and chloroprene (neoprene)
Load bearing	Polyurethane is superior to all
Bonding to metal	Natural, chloroprene, SBR and nitrile
Compression set	Silicone and fluoroelastomers
Electrical	Natural, SBR and silicone rubbers
Impact resistance	Natural and SBR
Abrasion resistance	Natural, chloroprene, SBR and nitrile
Tear resistance	Natural rubber
Cut growth	Natural rubber
Radiation resistance	Polyurethane best
Weather resistance	Neoprene, silicone, fluoroelastomers and polyacrylates
Oxidation	Silicone, fluoroelastomers and polyacrylates
Ozone resistance	Neoprene, silicone, fluoroelastomers and polyacrylates
Grease and oil	Neoprene, nitrile and fluoroelastomers
Water resistance	Natural, chloroprene, SBR and nitrile

Latex rubber can be poured like polyurethanes but is generally only poured in thin sheets. Even thicker sheets like Linatex® require a very long cure time. Pouring is the method of choice for cast polyurethanes. Normally rubbers require a compression molding press or injection molding machine to produce parts. This has high capital and mold costs.

9.3.4   Limitations of Polyurethanes

Polyurethanes can be used over a temperature of −40 °C to a maximum of 120 °C. The normal working range has a lower maximum temperature of 70 °C.

Hydrolysis is a major problem for polyurethanes even with the right choice of polyurethane and the use of antihydrolysis agents such as polycarbodiimides. Care and experience are needed in these conditions.

Certain chemicals such as ketones (acetone and MEK), polar solvents and concentrated acids, affect polyurethanes badly.

A major consideration in all dynamic applications is the buildup of heat. The low conductivity of the polyurethanes does not allow for rapid removal of heat. The design of the part and the grade choice is very important. The part must not be allowed to deflect too much and a heat sink must be provided. It is desirable to keep the cross-section of the polyurethane at a minimum.

9.4   Polyurethane Selection Criteria

Selecting a polyurethane for any application consists of two stages:

(1) Properties Required for Application

Every application needs to be evaluated on its merits for the range of properties required. A database of previous successful applications is a distinct advantage. An analysis of the main attributes of an application is needed, for example, compression, shear, static, dynamic or wear.

Many of these factors are not on the supplier's data sheets and must be obtained from the supplier or by experimentation.

The overall environmental conditions must also be evaluated. Some of the information may be hard to obtain from the customer due to commercial secrecy. Knowledge of the general type of application needs to be known, for example, a simple defoaming agent may be let down in diesel which has a moderate effect on polyurethanes. The effect of the overall concentration then needs to be evaluated.

Changes in temperature are an important limitation. The temperature may be from the process stream or it may be generated internally by dynamic work.

(2) Processing of the Polyurethane System

There are several factors that must to be taken into account when the exact grade of material is chosen.

The first is the pot life of the system. The pot life controls the time it takes to fill the mold, allowing any entrapped bubbles to rise and to gel off so that it can be placed in the curing oven. This represents a large portion of the physical cycle time.

Very hard polyurethanes gel rapidly (<one minute). Hand casting should be carried out at as low a temperature as possible to allow for complete filling of the mold. These polyurethanes are normally of a reasonably low viscosity. The pour hole in the mold must be as large as is practical to allow for the speedy addition of the mix. Care must be taken not to allow any folding of the mix in the mold.

PTMEG-based systems are normally much more viscous than PPG-based materials. They must be processed at as high a temperature as possible (at the upper limit of the manufacturer's range). The temperature however must not be so high as to cause problems in the chemistry during curing.

MDI-based systems are often slow in setting up. The addition of a catalyst may be needed to speed up the reaction. Catalysts such as tin salts (Polycat®[1]

---

[1] Air Products and Chemicals, Allentown, Pennsylvania.

T12 or equivalent) or an amine-based catalyst (Polycat 33LV or equivalent) can be used. Care must also be taken in the demolding of MDI-based systems as they take longer to build up enough strength to be demolded. Insufficiently cured parts will break easily. The surface of the break looks like shattered glass.

9.4.1 Applications in Tension

The ultimate tensile strength and the modulus at various strains are always quoted in the technical literature. These are very good quality control values and give an indication of the type and nature of the prepolymer system used. These tests are also valuable in evaluating the chemical and aging effects on polyurethanes. Where polyurethanes are used in tension, the amount it is stretched is normally no more than 30%. The highest extension would be in the order of 100%.

Using polyurethanes as a spring is an application where there is a tension and a compression cycle. The ability of the polyurethane under stress-strain cycling to reach a steady state is important. The material must, after the first couple of cycles, reach equilibrium to give a constant spring rate. TDI- ether-based materials are best suited for shock mounts or die springs. Applications such as sprockets need a tougher material such as a MDI-ester-based material.

Polyurethanes do not need fillers to give the hardness required as do rubbers. This means that they will keep their resiliency properties over a larger temperature range than conventional rubbers.

In all cases where polyurethanes are used under tension, the part must be designed to fail in a safe mode. Any internal defects can severely affect the life of the spring as failure often develops around these points.

9.4.2 Load-Bearing

Polyurethanes have very good load-bearing properties under compression and this is used in a number of applications. The ability to take the load and then to recover is superior to all other plastics. In all consideration it must be remembered that polyurethanes are virtually incompressible, that is, they change shape but keep the same volume. The parts must be designed so that the bulge does not become too large and put undue stress on the edges.

Examples of successful applications include wheels, tires, cutting and feed rolls, and metal forming pads.

The shape factor is an important consideration in the response to any applied load. Shape factor is defined as the ratio of the area of one loaded surface to the total of the unloaded surface that is free to bulge. The ability of the part to move when placed under load is important. If the surfaces are bonded to metal plates, the compressive stress to the compressive strain relationship is quite different.

Figure 9.1 illustrates the load bearing of a series of polyurethanes compared to SBR and neoprene rubber compounds.

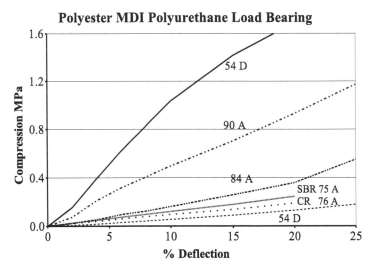

FIGURE 9.1
Polyester versus rubber load bearing.

Wheels

Polyurethanes are used in a variety of wheels, from small wheels in machinery to large wheels such as forklift and rolling wheels on mills.

The high degree of natural hardness of polyurethanes together with its excellent dynamic properties make it ideal for load-bearing wheels with the heat dissipation being the controlling factor. This limits the speed and time that the wheels can work. The very high resiliency grades of MDI-based systems makes the use of polyurethane a very good material for rollerskate wheels and similar applications.

In applications using polyurethane wheels, the normally used compression of the polyurethane is 10% of its original thickness. This can be obtained by adjustments to the hardness grade and compression modulus. Polyurethanes in the hardness range of 90 A to 50 D normally have enough load-bearing potential for applications in wheels. See Figure 9.2. The next factor to be considered is the heat buildup in the tire. Polyurethanes with a high resilience (low tan δ) are the most suited. Other properties that must be optimized are nick resistance and abrasive wear.

There are two normally quoted formulas for the calculation of wheel parameters. The first is the classical formula that is still quoted in imperial

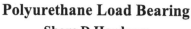

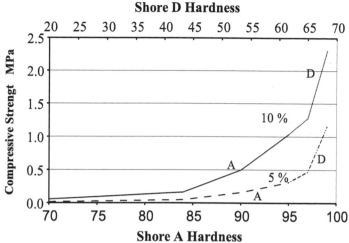

FIGURE 9.2
Polyurethane load bearing for wheels.

units:

$$U = \left[\frac{(0.75W(b-a))}{(ES(8b)^{1/2})} \right]^{2/3}$$

where
    U = deflection in inches
    W = load in pounds
    E = compression modulus in psi
    a = inside radius
    b = outside radius of polyurethane in inches
    S = actual width in inches

The value of U must be less than 10% of the thickness of the polyurethane $(b-a)$. This, as in all calculations, is the starting point. The suitability of the selected grade should be evaluated in either real or simulated tests. The second formula quoted is that of the European Tire and Rim Technical Organization. This is used for polyurethanes in the hardness range of 92 A to 95 A.

$$L = k \times 0.238 \times 10^{-8} \times (628 - 2t)^2 \times D \times W$$

where
    L = load in kilograms
    k = speed/use rating factor
    t = polyurethane thickness in millimeters
    D = outside diameter of polyurethane in millimeters
    W = width of polyurethane in millimeters

The measurements are illustrated in Figure 9.3.

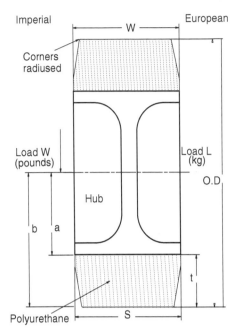

FIGURE 9.3
Wheel deflection formula.

The k factor varies from 115 to 70 depending on the speed and if any side forces are applied to the wheel. For example, at 10 kph, the load wheel of a forklift has a k factor of 100. If the speed is increased to 16 kph, the factor drops to 85. The steering wheel has a lower factor of 70 at 10 kph.

The heat generated in the tire must be able to be dissipated at below the maximum temperature of the bonding agent (approximately 115 °C). If the tire overheats and decomposes, it will be from the inside of the wheel.

Warning

Care must be taken in using polyurethane wheels in areas that are smooth and wet as the coefficient of friction is very low, and stopping and steering can be a problem.

9.4.3   Applications in Shear

In certain applications there are not only compressive loads but also a stress function (Figure 9.4).

Even though the polyurethanes have very good stress (tensile strength) properties, they deflect more under stress than compression. To overcome this problem the materials need to be bonded to metal plates to ensure that they do not slide out of place. The limiting factor now becomes the bond strength.

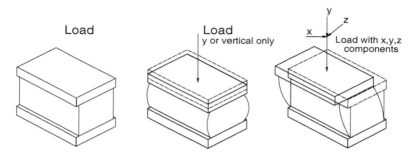

FIGURE 9.4
Parts under compression and shear.

The standard 90°peel test shows that the material breaks before the bond fails. In this situation the peeling component is nearer to 180°and the stress is more on the bond line (Figure 9.5).

The bonding material traditionally is a thermoplastic and it softens as the temperature increases. The bond strength decreases as the material becomes warmer. The shear component in a compression mount is often a result of misaligned components in a system. Differential thermal expansion of a steel girder compared to a concrete base will also give this effect.

9.4.4 Wear Resistance

The most important factor in any application is to determine the nature of the wear. Is the wear abrasive or is it erosive? When polyurethanes are used in a dry screening application, the particles move through the air and are not blown along in a current of air. This application is abrasive wear. In a dry cyclone where the air blows the particles, it is an erosive wear situation. The basic layout of a cyclone is given in Figure 9.6.

The centrifugal effects on impeller parts must be taken into account when using elastomers. Soft elastomers will tend to expand slightly when the speed is too high. It is advisable to reinforce the polyurethane part. The flow patterns in a pump are very complex and need to be evaluated very carefully. Incorrectly adjusted clearances can lead to turbulent flow and unexpected wear patterns. High-resiliency polyurethanes with hardness between 82 and 92 A prove to be satisfactory.

In an industrial situation the following factors affect the choice of material:

- Size and shape of the solids
- Velocity of the particle

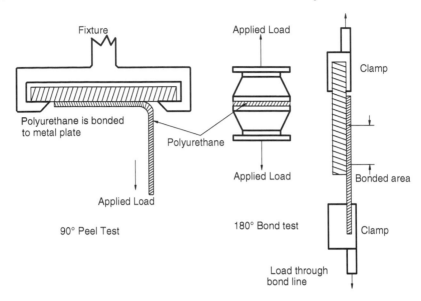

FIGURE 9.5
Polyurethane bond strength testing.

- Angle of impingement
- Operating temperature
- Chemical composition of process medium
- Flexibility of polyurethane

The chemical composition and temperature of the process medium greatly influences the choice of material. The acidity or alkalinity together with the standard operating and maximum temperature must be taken into account when the material is selected. The toughness and excellent wear of the polyesters must be balanced against maximum working temperature, ease of processing and the cost.

The presence of a reinforcing will increase the stiffness in situations such as screens and pump parts. This will also act as a heat sink. In cyclones there are a number of approaches:

- All solid polyurethane
- Polyurethane liners in metal casing
- Polyurethane liners in composite (DMC) casing

The under- and overflow spigots are normally made from a harder material and the angle of wear is very much more sliding and almost abrasive in nature.

9.4.5 Vibration Damping

The hysteresis property of any polyurethane consists of two components, namely the spring and the viscous component. The viscous component is re-

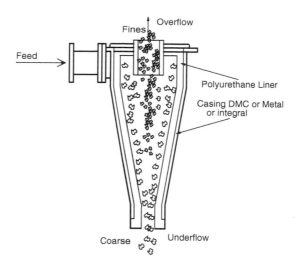

FIGURE 9.6
Basic cyclone layout.

sponsible for the absorption of the force and the conversion of the energy into heat.

Any deflections must be within the limits where Hooke's Law is obeyed (i.e., only 1–10%). The ratio between the forced and natural frequencies needs to be determined. The natural frequency is a function of the static deflection of the system. The damping ratio of the polyurethane must also be known. This can vary from 0.05 for highly resilient materials to 0.15 for low-resiliency materials. To obtain damping, the forced-to-natural damping ratio must be greater than 1.4.

Polyurethanes can be used to develop vibration isolation pads to isolate movements generated in the machinery.

Absorption of acoustic vibrations can also be done with polyurethanes. This finds application in submarine coatings for certain temperature conditions.

9.4.6 Electrical

Polyurethanes are not suited as an insulation material in mains voltage situations due to the ability of the polyurethane to absorb moisture and lower its resistivity. Polyurethanes alone or in a combination with an epoxy can be used as a potting material.

Polyurethanes can be modified to be semiconductive throughout the whole volume of the material, which allows them to be used in applications where generated static electricity has to be dissipated. Typical examples of these are textiles or paper passing between rollers at speeds where an electrical charge can develop, leading to poor handling and the risk of electrical sparks. Grains flowing down a polyurethane chute can also build up an electrostatic charge. As there is normally a large amount of organic dust in the area, any discharge could potentially cause an explosion.

Antistatic polyurethanes find applications in electronic and business machines when there are moving parts that can generate static. They can also be used where benchtops need to be kept at zero electrical potential.

9.5   Design

9.5.1   Bonding

Bonding of polyurethane to metal is needed to keep the part in the correct position and to provide a continuous layer between the reinforcing material and polyurethane.

Large flat objects require holes in the reinforcing to allow the air to escape when the part is poured. Holding the part at an angle also helps. The result of not getting rid of all the air is to have large areas with no bond. On heating over time, moisture can permeate and blisters may be formed. The formation of a blister is shown in Figure 9.7.

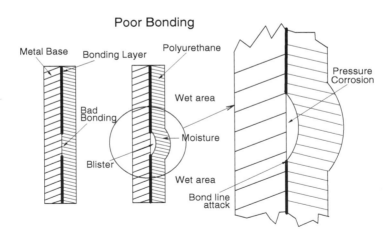

FIGURE 9.7
Poor bonding causing blisters.

When parts are to be used in tension, the polyurethane should be bonded onto a tag rather than a flat plate. This helps eliminate stresses on the bonded surface.

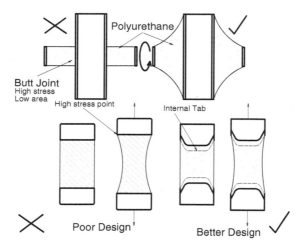

FIGURE 9.8
Reduction in stress using fillets.

9.5.2 Fillets

The reduction in potential stress points in any part is very important. The use of fillets in the junction between the polyurethane and the reinforcing is important in limiting stress concentration. Right-angled joints are not desirable. The use of fillets to reduce stress is shown in Figure 9.8.

9.5.3 Shape

The correct choice of the shape factor in polyurethane parts can help the bulging of parts under compression. This is shown in Figure 9.9. A cylindrical part with the same loaded area will deflect less than a rectangular part under the same load. Any surface that is bulged due to compressive forces will be more prone to failure due to nicking or fatigue. A simple solution is to design the part so that the material under no load has a slight "waist." On compression, the material will give an item with more parallel sides and hence less stress on the surfaces.

Radius or chamfer the outermost edges of wheels and bushes. Square edges will bulge out under load and tend to break off. This will give more sites for further damage.

To increase the vertical stiffness of any load-bearing component, multiple layers of steel plates can be inserted into the molding. This will increase the vertical stiffness while having no influence on the shear properties. There will be more bulges of smaller magnitude than if a single unit is poured.

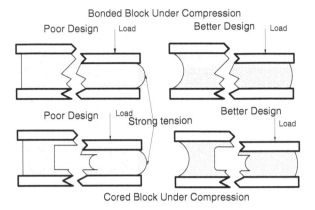

FIGURE 9.9
Bonded and cored blocks under compression.

9.6 Summary

Polyurethanes find many different uses in a wide variety of industries. Their applications are only limited by the ability to use the right grade with an appropriate design change. The major limiting factors are the operating temperature range and any specific chemical attack.

Appendix G lists a number of common applications of polyurethanes. This list is by no means complete as there are many unpublished applications.

10

Tools for Evaluation

10.1 Introduction

There are five main points in the evaluation of polyurethanes:

1. Verification of selected grade
2. Quality control tests
3. Type tests
4. Prototype and service tests
5. Investigative research

Full chemical and physical testing is often outside the scope of many processors of polyurethanes. Firms will mainly be equipped to carry out tests relating to their main product line. Many of the other tests may be carried out by specialist consulting laboratories or by a local university with an interest in the field that the article will be used in.

Wherever possible, testing should be carried out in accordance with an approved standard. These include the standards issued by the International Standards Organization (ISO), the American Society for Testing and Materials (ASTM) and the Deutsches Institut für Normung (DIN). There are many other standards of international repute such as the NF, JIS, SAC and BS standards. Most countries have their own standards. These are adjusted to suit local requirements but are based on the main international standards. Evaluations carried out to these standards in a laboratory compliant with ISO 9001 quality standard carry more weight than those tested under other conditions. There are other standards such as aviation and maritime specifications that are also recognized in their field. In all cases the most recent issue of the standard or specification must be used.

International Standardization

The current trend is for many nations to endorse the International Standards Organization (ISO) specification as the national specification.
For example,
BS 903-A1:1996 Physical testing of rubber Determination of density has been superseded by
BS ISO 2781:2008+A1:2010

These standard tests do not cover all situations. Simulated service tests or in-service evaluations should also be carried out. In-service tests should only be carried out if the potential for a positive result is very high or if requested by the customer.

10.2   Verification of Selected Grade

There are two main factors to be checked, namely:

1. Is it the right material for the application?
2. Are there any processing changes to be made?

These two factors are interrelated as the material needs to be processed before the evaluation can be carried out. Successful casting may require adjustments to the mold, processing conditions, and post-curing operations. Dimensional testing should be carried out at this point to check the need for any mold alterations that may be required.

Prior to any production for complex and critical applications, theoretical calculations should be carried out to assess the properties of the selected grade. Computerized mathematical techniques such as finite element analysis can be carried out to determine potential stress points. There are also mold filling programs that can be used. Both these methods are expensive and need specialized staff.

The evaluation process consists of several stages, with reassessment if the initial results are not satisfactory. The molding process or the grade of polyurethane may need to be changed. This may have to be repeated when service tests are carried out.

10.3   Quality Control Tests

10.3.1   Weighing Equipment

All scales must be kept clean and free from any buildup of polyurethane raw materials. The level and zero must be checked daily. A standard "check weight" must also be available in the normal weight range. The scales must be checked annually by a suitably qualified technician.

10.3.2   Temperature

The correct indication of temperature is important and the probes must be checked on a regular basis. Temperature indicators should have an internal check of the continuity of the probe's wires. Any breakage should be indicated on the display. The probe's sheath should be made from a material that is not attacked by any vapors that may be present. Both RTD and "K" type probes have been shown to be satisfactory.

Prepolymers heated in a microwave oven, even using a rotating table, will have uneven temperatures and the material needs to be mixed well prior to taking a temperature reading.

There are two main ways of calibrating the temperature indicating system. The first is to use a specialized test box that will either give a known voltage or resistance to the indicator. The output from the box corresponds to a known temperature. The output from the probe can also be checked. The second method is to check the melting point of ice, which should give zero on the scale. The boiling point of water is very close to 100 °C varying depending on the atmospheric pressure. The probe should be placed in steam at atmospheric pressure. There are tables available that correlate the atmospheric pressure to the boiling point of water.

The location of the temperature probes is very important to obtain the correct temperature of the curing ovens. They must be in the air, away from direct heat from any elements.

Measuring the temperature of cured polyurethanes requires a special technique. Polyurethanes are a poor conductor of heat and the probe will locally heat or cool the surface of the material. Once a temperature is obtained, the probe must be quickly moved to a second and third spot to obtain the correct temperature. The use of a noncontact thermometer (such as an infrared thermometer) can eliminate this problem. The instrument must be of the correct range and resolving power.

10.3.3   Linear Dimensions

Before carrying out any measurements, the material must be at ambient temperature and stabilized at a relative humidity of approximately 50 percent. It is realized that measurements may need to be taken during processing such as machining. In this case the expansion of the polyurethane by heat needs to be taken into account.

The very hard materials can be checked using the standard pressure settings. For softer materials this will cause undue distortions. The micrometer or vernier caliper needs to be gently passed over the item. The faces of the instrument must just touch the surface. This can be felt through the instrument as a very mild resistance. The measurements will represent the high points on the surface.

The micrometers and calipers must be zero checked before use and regularly checked against standard test blocks.

Standard tests for taking dimensions include:

Method	Details
ISO 23529:2010	Rubber–General procedures for preparing and conditioning test pieces for physical test methods
ASTM D3767 - 03(2014)	Standard Practice for Rubber-Measurement of Dimensions
BS ISO 23529:2010	Rubber–General procedures for preparing and conditioning test pieces for physical test methods

10.3.4 Density

The density, or more correctly the apparent density, of polyurethanes can be determined in a number of ways. Density is defined as the weight of the material in air divided by its volume.

The determination of the volume is the most complicated. The sample of polyurethane must be free from all internal bubbles or other voids. If the sample is of uniform shape, the volume can be determined geometrically by measuring its dimensions and then calculating its volume.

If the sample is irregular in shape, the traditional Archimedes principle must be employed. The sample is first weighed in air on a laboratory scale and then suspended by a fine thread in water. The density is given by:

$$\text{Sample density} = \frac{(M1 \times Qf)}{(M1 - M2)}$$

$M1$ = weight of sample in air
$M2$ = weight of sample in liquid
Qf = density of liquid, if water is used, Qf =1.0

Standard test for the determination of density include:

Method	Details
ISO 2781:2008(Amd1):2010	Rubber, vulcanized or thermoplastic – Determination of density
ASTM D1817-05(2011)	Standard Test Method for Rubber Chemicals – Density

There are other methods for the determination of density. These include the specific gravity balance and the use of a pycnometer (weighing bottle).

10.3.5 Hardness

Although not greatly influenced by the mixing ratio of prepolymer and curatives, hardness is one of the most frequently specified and carried out tests. The two main hand-held units are the Shore A and D range durometers.

The construction constraints and the test methods are detailed in the following specifications:

Method	Details
ISO 7619-1:2010	Rubber, vulcanized or thermoplastic – Determination of indentation hardness – Part 1: Durometer method (Shore hardness)
ASTM D2240-05(2010)	Standard Test Method for Rubber Property – Durometer Hardness
DIN ISO 7619-1	Rubber, vulcanized or thermoplastic – Determination of indentation hardness – Part 1: Durometer method (Shore hardness) (ISO 7619-1:2010)
ISO 7619-2:2010	Physical testing of rubber. Determination of indentation hardness by means of pocket hardness meters

For the most accurate results, dead-load instruments should be used. Traditionally all the instruments have been analog but currently both analog and digital units are available.

The most important features of the instruments are:

- Size and shape of the indentor
- Distance the indentor protrudes from the base
- Characteristics of the spring

The A scale instrument has a frustoconical indentor while the D scale has a pointed conical indentor. The scales overlap but the stresses on the material are different. There are also micro units available for measurements on small pieces of material, including "O"-rings.

Figure 10.1 illustrates the method of operation of the durometer hardness testers.

New electronic units claim to have an accuracy of ± 1 unit, which is much better than the previous ± 3 units. An experienced operator can obtain more consistent results than the ± 3 units range quoted. Results may vary from operator to operator.

A check spring is supplied with the durometer and can be used for day-to-day checks. A series of standard rubber blocks can also be purchased for checking over the full range. The blocks need to be rechecked every two years. Durometer suppliers also offer a recalibration service.

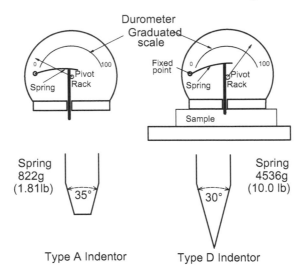

FIGURE 10.1
Durometer: basic layout.

10.3.6   Curative Level

Certain curatives, such as those containing either chlorine or sulfur, can be
tested using x-ray fluorescence (XRF) techniques. The principle of XRF is
that the electrons in the atoms become excited when exposed to the x-rays.
As the electrons move in the atoms, they emit x-ray light (photons). The
wavelength is characteristic of the element and the intensity proportional to
the concentrations in the sample. Elements with an atomic number greater
than 15 (magnesium) can be determined by this method. Standards have to
be made for the different levels and types of curative.

As the method is independent of how the atom is attached, chlorine containing plasticizers such as TCP will interfere with the result if MOCA is
used.

There are some simple units that use a radioactive source instead of an
x-ray generator. These have problems with handling and storing the source,
as well as disposal of the spent source after a number of years.

10.3.7   Surface Porosity (Spark Test)

When a thin layer of polyurethane is cast over a metal reinforcing, the thickness of the polyurethane must be even and sufficient to prevent moisture
from getting to the reinforcing and causing corrosion. A potential cause of
problems is if an air bubble is trapped between the reinforcing and the cast
polyurethane.

The entire thickness of the polyurethane over the reinforcing can be measured using an ultrasonic gauge. The correct head and sound velocity will have to be used as polyurethanes require different conditions than metals.

An alternative method is to use high-voltage spark testing (500 V to 30 kV) either in the DC or AC mode.

There are several methods for carrying out these tests:

Method	Details
ISO 2746:1998	Vitreous and porcelain enamels – Enameled articles for service under highly corrosive conditions – High voltage test
DIN 55670	Paints and varnishes – Method for testing paint coatings for pores and cracks using high-voltage
ASTM D4787-13	Standard Practice for Continuity Verification of Liquid or Sheet Linings Applied to Concrete Substrates
ASTM A742 / A742M-13	Standard Specification for Steel Sheet, Metallic Coated and Polymer Precoated for Corrugated Steel Pipe

The voltage in the units needs to be set to enable the detection of a defect but it must not be so strong that it burns a hole in the polyurethane. Either a brush or wand can be used to check the continuity of the surface. This type of test can be used with different fittings to check the interiors of items such as pipes.

10.4 Type Tests

Type tests of polyurethanes are used to evaluate polyurethanes using standardized methods. The tests are a characteristic of the polyurethane being evaluated but do not represent actual field conditions.

The samples need to be brought to controlled test conditions before the test is carried out. If they were used as a quality control test, the product in many cases would be at the customer before the test is completed. A typical example is that tests such as the tensile tests require a week to settle down before the test is carried out after the 18-hour cure cycle. In many cases an approximate value is obtained by testing as soon as the sample has reached room temperature after it has had its full cure.

10.4.1 Tensile Stress-Strain

Tensile properties are used in the evaluation of batches of material as well as in determining the aging effects on polyurethane. Currently the time taken for the preparation and stabilization of the samples does not make this test ideal for a batch-to-batch real-time quality control tool.

Testing is normally carried out to one of the following standards:

Method	Details
ISO 37:2011	Rubber, vulcanized or thermoplastic – Determination of tensile stress-strain properties
ASTM D412-06a(2013)	Standard Test Methods for Vulcanized Rubber and Thermoplastic ElastomersTension
DIN 53504	Testing of rubber – Determination of tensile strength at break, tensile stress at yield, elongation at break and stress values in a tensile test

There are a number of different terms for the strength of the polyurethanes. These include:

- Tensile strength
- Breaking stress
- Breaking load
- Ultimate tensile strength

Tensile strength is the highest tensile stress reached before the sample breaks. Conventionally the result is expressed as the force required per unit area. The area is taken as the original cross-sectional area. The sample is dumbbell shaped and the area is measured in the waist of the sample.

The preparation of the sample is important. There should be no gas bubbles or other imperfections in the test sheet. The cutting knife must be sharp and free from any nicks. The sides of the sample must not be concave or have any fine ridges. Tensile samples from hard grades may be milled out.

Clamps used to hold the sample must be self-tightening as the polyurethane shrinks under load. Any traces of mold release act as a lubricant and make the holding of the sample difficult. The flat tag ends of the dumbbell can be slightly roughened. Pneumatic clamps can help keep the samples in place.

As the samples are dumbbell shaped, the movement of the crosshead of the tensile machine cannot be used to determine the elongation. Extensometers of various design are used to measure the elongation in the waist area of the sample. One method is to have two clamps set to the desired distance apart. As the sample stretches, the movement is measured by a system of counterweights and cords. Another method is to use light/laser beams reflecting from pieces of reflective tape on the sample.

The "elongation at break" or "ultimate elongation" is the elongation expressed as a percentage at the point of rupture.

The modulus (or tensile stress) at various elongations is the load at the elongation divided by the original cross-section. This is expressed in terms of force per unit area (MPa).The normally used elongation values are 50%, 100%, 200% and 300%.

When using the tensile strength or modulus at a certain elongation to investigate change in properties, it is better to use relative changes to the ultimate values. A material that has changed from 28 MPa down to 15 MPa has suffered more damage than one that has changed from 21 MPa to 14 MPa (54% and 67% retained).

10.4.2   Set — Tension and Compression

Tension Set

The test involves stretching the sample to a constant strain (generally 100% or 25%, 50%, 200% or 300%). The sample is held at the desired strain for a fixed period before the load is removed and the sample allowed to stabilize. The increase in length of the markers is the permanent set.

The methods are detailed in:

Method	Details
ISO 2285:2013	Rubber, vulcanized or thermoplastic – Determination of tension set under constant elongation, and of tension set, elongation and creep under constant tensile load
ASTM D412 - 06a(2013)	Standard Test Methods for Vulcanized Rubber and Thermoplastic Elastomers – Tension

The test consists of marking the desired length on the dumbbell, stretching the sample to the desired elongation and holding at that elongation for a set period of time. The time duration may be 24, 72 or 168 hours. The temperature may be ambient, 70, 85 or 100 °C. The test time starts 30 minutes after the sample is introduced and the final measurements made after 30 minutes recovery.

$$\text{Tension set} = \frac{(L1 - L0)}{(LS - L0)} \times 100$$

where
  $L0$ = original unstrained length
  $L1$ = reference length after recovery
  LS = strained reference length

Compression Set

Samples of polyurethane are clamped between two plates so that they are compressed by 25%. The samples may be tested at ambient or 70 °C. They are compressed for either 24 or 72 hours. After compression they are allowed to recover for 30 minutes and then remeasured.

The methods are detailed in:

Method	Details
ISO 815-1:2014	Rubber, vulcanized or thermoplastic – Determination of compression set – Part 1: At ambient or elevated temperatures
ISO 815-2:2014	Rubber, vulcanized or thermoplastic – Determination of compression set – Part 2: At low temperatures
ASTM D395-2014	Standard Test Methods for Rubber Property – Compression Set

If there is no recovery the compression set is 100% and with full recovery it will be 0%.

$$\text{Compression set at constant strain} = \frac{(TO - TR)}{(TO - TS)} \times 100$$

where
   TO = original thickness of piece
   TR = thickness after recovery
   TS = thickness of spacer

Figure 10.2 illustrates the sample sizes and basic test equipment for the compression set test. The main problem with the compression set results is that the time period of testing is so short in relation to the real-life situation; for example, a gasket seal may be in service for one or more years before it is removed. It is also found that if left for longer periods than the 30 minutes specified, the compression set becomes less.

10.4.3   Tear Strength

In tensile strength determinations the material has to completely break through the cross-section, whereas the tear strength test indicates the resistance to the propagation of a defect, such as a nick, in the polyurethane. The way that elastomers tear under different conditions has led to a number of different tests.

There are three main configurations for the tear test. In the first two a tensile force is applied to the sample. In one configuration the force is vertical to the plane of the sample and in the second it is at right angles. In the third style, an element of shear is also introduced. Figure 10.3 shows the basic configuration of the tear test pieces.

There are a number of different standard test methods. These include:

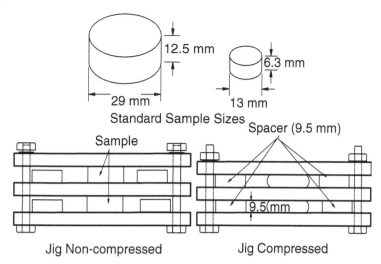

FIGURE 10.2
Compression set layout.

Method	Details
ISO 34-1:2010	Rubber, vulcanized or thermoplastic – Determination of tear strength – Part 1: Trouser, angle and crescent test pieces
ISO 34-2:2011	Rubber, vulcanized or thermoplastic – Determination of tear strength – Part 2: Small (Delft) test pieces
ASTM D624 - 00(2012)	Standard Test Method for Tear Strength of Conventional Vulcanized Rubber and Thermoplastic Elastomers

There are a number of different-shaped samples that are specified for carrying out the tear strength determination. The exact shape and test method used need to be specified for any values quoted.

The two most commonly specified tests are the Die C test and the "trouser tear" (ASTM D624) test. Both of these tests measure resistance to tear propagation. Other test shapes include the crescent tear with or without end tabs. This sample shape is often nicked. The "Delft" (ISO 34-2) sample with its internal nick is often used from samples taken from finished products. The exact size of the nick does not appear to be critical to the result.

The standard calculation is based on the force required for the tear (in Newtons) divided by the thickness of the sample in millimeters. When carrying out the trouser tear, the median force must be taken as the load varies during the test.

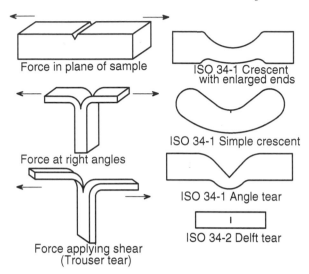

FIGURE 10.3
Tear strength test pieces.

In practice there are many factors that may cause a sample to fail. These include nicking under tension, impurities in the material or air bubbles.

10.4.4   Adhesion

Very strong bonding between polyurethane and metal is needed in items such as polyurethane mounts. When metal is used as a reinforcing or for location and stabilization, a very good bond is needed to form the bridge between the rigid metal and the elastic properties of the polyurethane. Ideally a test that could be performed on the finished product is desirable but in practice this is often not the case.

Air pockets or blisters can be detected using ultrasonic test methods. The correct head and frequency needs to be used as the speed of sound through elastomers is very different than through metals.

Peel Tests

Peel tests, or the "one plate" method, are commonly used. These are specified in the following standards:

Method	Details
ISO 813:2010	Rubber, vulcanized or thermoplastic – Determination of adhesion to a rigid substrate – 90 degree peel method
ASTM D429-14	Standard Test Methods for Rubber Property – Adhesion to Rigid Substrate

In peel tests the polyurethane is bonded to a metal strip. The bonded area is 25 by 25 mm. The sample is conditioned and an initial length of approximately 1.5 mm is stripped from the plate using a knife. The sample is mounted in a tensile strength machine and the polyurethane stripped from the plate. Figure 10.4 shows the sample configurations used. The adhesion is calculated by dividing

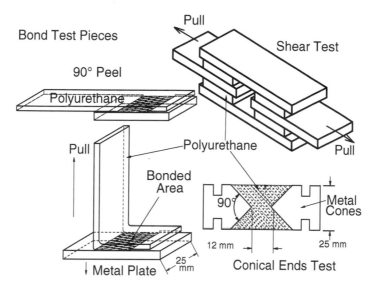

FIGURE 10.4
Bond test sample configuration.

the maximum force recorded by the sample width in millimeters. With the current bonding technology, the polyurethane will in most circumstances break before the sample starts to peel. This is referred to as a stock break (SB).

If the sample does peel off, the zone where the peeling takes place is recorded. If the peeling occurs between the metal plate and the bonding medium, the surface preparations needs to be examined. The bonding may fail between the bonding medium and the polyurethane. This could be caused by either poor activation of the bonding agent or contamination of the surface by mold release. If multilayers of bonding agents were used, there can be interlayer failure caused by contamination or improper activation.

It has been suggested that the test angle be changed from 90 to 45°to give more realistic and reproducible results. This has not met with official approval.

Tension Tests

There are two main types of tension tests in use, namely the "two plates" test and the "conical ends" test.

Method	Details
ISO 814:2011	Rubber, vulcanized or thermoplastic – Determination of adhesion to metal – Two-plate method
ISO 5600:2011	Rubber – Determination of adhesion to rigid materials using conical shaped parts
ASTM D429-14	Standard Test Methods for Rubber Property – Adhesion to Rigid Substrates (methods A and C)
DIN 53531-2	Determination of the adhesion of rubber to rigid materials using conical ended cylinders

Care must be taken to align the samples exactly vertically in the tensile strength machine, otherwise shearing is introduced into the mechanism.

In the conical test, a high concentration of stress is placed at the apex of the cones and any failure starts at this point.

10.4.5   Shear Tests

In this test multiple bond points are used and a quadruple bond test is performed. These tests are illustrated in Figure 10.5.

Shear tests are specified in the following specification:

Method	Details
ISO 1827:2011	Rubber, vulcanized or thermoplastic – Determination of shear modulus and adhesion to rigid plates – Quadruple-shear methods

The test pieces are similar to that used to determine shear modulus. The results are obtained by dividing the maximum force required by the total bonded area of one pair of samples.

All these tests are carried out under relatively slow conditions (50 mm/minute) and do not represent normal service conditions. There are no standard tests to simulate these relief conditions. Equipment is available to perform these dynamic tests.

These tests are useful as a screening test to evaluate new bonding agents and methods of surface preparation. It must be remembered that polyesters have a bond strength three times that of polyethers.

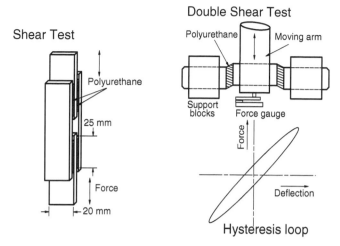

FIGURE 10.5
Double shear test.

10.4.6 Compression Modulus

Polyurethanes are more often used in compression or in the shear than in the tensile mode. There are a number of standard tests that may be used:

Method	Details
ISO 7743:2011	Rubber, vulcanized or thermoplastic – Determination of compression stress-strain properties
ASTM D575-91(2012)	Standard Test Methods for Rubber Properties in Compression

The test piece is a cylinder of 29 ±0.5 mm diameter by 12.5 ±0.5 mm high. The sample is compressed in a tensile strength machine. The sample must not be allowed to move on the compression anvils (fine emery paper is suggested between the anvil and sample).

The result of the third compression on the sample is taken. During the first two, the sample is compressed 5% more than the maximum to be recorded. It is recommended that the results be given graphically as compression strain versus compression stress.

In the compression modulus test the movement of the crosshead can be taken. Precautions must also be taken to prevent over-run of the crosshead by having stops installed to prevent damaging the load cell with an over-run.

10.4.7 Shear Modulus

Applications where there is compression will often have a shear component even if not by design.

The following method details the determination of shear modulus:

Method	Details
ISO 1827:2011	Rubber, vulcanized or thermoplastic – Determination of shear modulus and adhesion to rigid plates – Quadruple-shear methods

The ISO 1827 method has four blocks 4 mm thick by 20 mm wide and 25 mm long bonded to plates as shown in Figure 10.5. The samples are cycled five times in a tensile machine before the results are taken. A micrometer or the movement of the crosshead can be used to measure the strain. The speed of the machine is 25 ± 0.5 mm per minute.

The shear stress results are expressed in Pascals (N/m^2). The calculation is obtained by dividing the applied force indicated by the tensile machine by twice the actual bonded area of one block in square meters (nominally 20 x 25 x 10^{-6}). The shear strain is obtained by dividing half the actual deformation by the actual thickness of one block. All measurements must be in the same units.

10.4.8 Dynamic Mechanical Testing

Tests such as the tensile strength and tear strength tests evaluate polyurethanes to destruction. When polyurethanes are used in a practical situation, the aim is for them to have as long a life as needed in the application. Stress, strains and shear are applied to the polyurethane at various frequencies and at different temperatures. There may also be dynamic variations on top of a static load, for example, vibrations on a loaded isolation pad.

ISO 2856 and ASTM D2231 outline general requirements for dynamic testing but with no descriptions of actual method or equipment. The tests are divided into three main groups:

1. Test methods where there is free vibration applied and then allowed to decay.

2. Tests where the vibrations are maintained from an external source. The frequency can be adjusted to be either resonant or nonresonant.

3. Propagation of either continuous or pulsed waves below the kHz or MHz range.

There are a number of methods for this test and they include:

Method	Details
ISO 6721-1:2011	Plastics – Determination of dynamic mechanical properties – Part 1: General principles
ISO 6721-2:2008	Plastics – Determination of dynamic mechanical properties – Part 2: Torsion-pendulum method
ASTM D5992-96(2011)	Standard Guide for Dynamic Testing of Vulcanized Rubber and Rubber-Like Materials Using Vibratory Methods

These tests form the classical methods of testing but the use of Dynamic Mechanical Analyzers is becoming more popular, especially those that can operate from subambient temperatures to 150 °C and above.

10.4.9 Resilience

Resilience is one of the most common rebound tests that are carried out. It is fundamentally a deformation of the material for half a cycle. The rebound resilience basically is the ratio of the indentor after to before impact expressed as a percentage.

There are a number of methods for this test and they include:

Method	Details
ISO 4662:2009	Rubber, vulcanized or thermoplastic – Determination of rebound resilience
ASTM D1054-02(2007)	Standard Test Method for Rubber Property-Resilience Using a Goodyear-Healey Rebound Pendulum (Withdrawn 2010, no replacement)
ASTM D2632-14	Standard Test Method for Rubber Property Resilience by Vertical Rebound
DIN 53512	Testing of rubber – Determination of rebound resilience (Schob pendulum)

The tests all have different test conditions and only results from the same type of instrument can be compared. The instruments are all simple in design but with different degrees of complexity.

ISO 4662 details testing by the pendulum method. There are a number of different machines that conform to the specifications for impacting mass, velocity and the apparent strain energy density. The types of suitable instruments mentioned in the specification include the Lüpke, Schob and Zerbini pendulums. Some modern instruments have electronic outputs giving either digital or printed results. Falling plunger and ball machines are generally not popular in Europe but one design is specified in ASTM D2632. ASTM 1054 describes the Goodyear-Healey rebound tester. Figure 10.6 shows the principles of some of the tests.

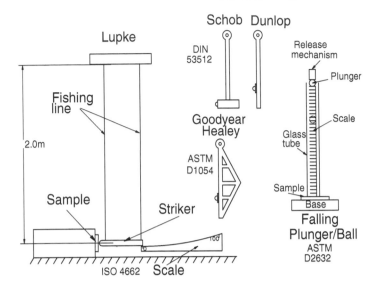

FIGURE 10.6
Resilience test methods.

Forced Vibrations

Tests based on forced vibrations are preferred when parts are being designed. The relationships are still very complex and only approximate. If experimental tests can be carried out close to actual service conditions, better results can be obtained.

Forced vibration testing is detailed in:

Method	Details
ISO 4664-1:2011	Rubber, vulcanized or thermoplastic – Determination of dynamic properties – Part 1: General guidance
ISO 4664-2:2006	Rubber, vulcanized or thermoplastic – Determination of dynamic properties – Part 2: Torsion pendulum methods at low frequencies
DIN 53513	Determination of the viscoelastic properties of elastomers on exposure to forced vibration at non-resonant frequencies

There are a number of different types of machines available to perform the tests based on either a mechanical, hydraulic or electromagnetic method to obtain the dynamic movement in the samples. The frequency of the applied force can vary between 2 Hz and 10,000 Hz using an electromagnetic vibrator.

A typical machine is the Rhiovibron^{TM}. In the test, the samples are bonded onto the fixture either during curing or by using cyanoacrylate glue.

From the results obtained when used in a dynamic stress strain mode, a hysteresis graph can be generated. As the machines were developed and computerized results obtained, a new group of instruments was developed, namely the dynamic mechanical analyzers.

Free Vibration Testing

The torsion pendulum is detailed in ISO 4664-2 where one end of the sample is fixed and the other can rotate freely. Frequencies of vibration of between 0.1 and 10 Hz are possible with this method. An older method specified under ASTM D945 is the Yerzley oscillograph where a beam can oscillate vertically. The die-off of the vibrations is recorded on a drum. Resilience results can be calculated.

10.4.10   Dynamic Mechanical Analysis (DMA)

Polyurethanes have a combination of elastic and viscous properties that can be explained in standard engineering terms using DMA methods. Information can be obtained on the properties of polyurethanes that relate to the storage and dissipation of energy applied during use.

The equipment can operate in several different modes, over various frequencies and over a wide temperature range. Typical modes that can be used are:

- Tension
- Compression
- Three-point bend
- Single or dual cantilever
- Shear sandwich (for very viscous materials)
- Submersed sample cups

The manner in which these modes are carried out can further be modified by some of the following methods:

- Multistress-strain applications
- Multifrequencies
- Creep/stress relaxation
- Constant strain
- Controlled force/strain

The instruments can operate from $-150\,^{\circ}\mathrm{C}$ to well above the melting point of polyurethanes.

Samples that are used must be free from any defects such as bubbles and must be uniform in size and measured accurately. The size of the samples is reasonably small, approximately 3 mm thick by 12 mm wide and 50 mm long. The exact size will depend on the instruments and the test being carried out.

The basic results that are generated are:

- Storage modulus (energy stored):

  - E' Stretching
  - G' Twisting or torsional

- Loss modulus (energy lost - generally as heat):

  - E'' Stretching
  - G'' Twisting or torsional

- Tan delta (tan δ)

Tan delta is defined as

$$\tan \delta = E''/E' = G''/G'$$

Other outputs from a typical instrument include complex modulus, creep compliance, relaxation modulus, sample stiffness and all the basic data including instrument parameters.

The operation of the instrument involves the selection of the correct mode and the frequency and method of application. The sample is cooled to the start temperature and allowed to stabilize. A small deformation is applied and the results obtained. The temperature is raised and the process is repeated until the final temperature is reached.

The uses of such a technique include:

- Studies of the effect of changes in chemistry on the properties of the polyurethane
- Determination of the lower and upper useful temperatures of the polyurethane
- Provision of dynamic engineering properties
- Creep determination
- Coefficient of expansion

These instruments are high precision with state-of-the-art force and position measurement technology. Modern computer power enables the storage of data obtained during the test as well as its mathematical manipulation and presentation. Interpretative skills are required to obtain the best results from the techniques used. Comparison with results obtained in the field should always be made. There are a number of methods for this test and they include:

Method	Details
ISO 6721-7:1996	Plastics – Determination of dynamic mechanical properties – Part 7: Torsional vibration – Non-resonance method
ASTM D4065-12	Standard Practice for Plastics: Dynamic Mechanical Properties: Determination and Report of Procedures
ASTM D5026-06(2014)	Standard Test Method for Plastics: Dynamic Mechanical Properties: In Tension
ASTM D5279-13	Standard Test Method for Plastics: Dynamic Mechanical Properties: In Torsion

10.4.11 Electrical Properties

Polyurethanes can be used in applications where electrical properties are important. They are not normally used for high-voltage insulation. Polyurethanes are often used directly or in combination with epoxies for encapsulation. The addition of antistatic agents to polyurethane gives a product with the correct electrical properties while retaining the excellent wear needed for a number of roller-type applications.

Testing of polyurethanes for their electrical properties due to the voltages required must be carried out using properly designed equipment. The electrical tests that are normally carried out are resistivity, insulation resistance, electric strength, tracking resistance, power factor and permittivity.

Any application involving mains power must be approved by the local and state authorities before use.

Polyurethanes with antistatic properties are suitable for use where static electrical charges must be dissipated. The polyurethane must be compounded to provide the antistatic properties throughout the whole bulk of the material. The antistatic agent, if a liquid, must not migrate to the surface readily. Surface coating the item is not desirable as the effect is only temporary. Volume and surface resistivity tests for conductive and antistatic materials include:

Method	Details
ISO 1853:2011	Conducting and dissipative rubbers, vulcanized or thermoplastic – Measurement of resistivity
ISO 2878:2011	Rubber, vulcanized or thermoplastic – Antistatic and conductive products – Determination of electrical resistance
ASTM D991-89(2014)	Standard Test Method for Rubber Property – Volume Resistivity of Electrically Conductive and Antistatic Product

10.4.12   Environmental Resistance

Weathering Resistance

The Panama weathering study [6] has shown that polyurethanes are very resistant to the long-term effects of weather. There are a number of short-term effects that must be noted. Aromatic polyurethanes will yellow and become dark on exposure to light. These changes can have a drastic effect on the color of pigmented polyurethanes. Light colors should be avoided.

The effect of ultraviolet (UV) light on aliphatic clear polyurethanes or aromatic polyurethanes that have had UV stabilizers added can be evaluated using a Weather-Ometer^{TM}. The tests should be carried out in accordance with:

Method	Details
ISO 4665:2006	Rubber, vulcanized or thermoplastic – Resistance to weathering
ASTM D750 -12	Standard Practice for Rubber Deterioration Using Artificial Weathering Apparatus

There is often some surface crazing and chalking found in some samples when exposed to the weather.

In evaluating polyurethanes in weather exposure situations, care must be taken to consider the bulk of the material and not just the surface. The surface may be affected by the weathering but this may only be very shallow and not represented by the bulk of the material.

Chemical Resistance

Water, due to its molecular size and polarity, is absorbed into polyurethane and an equilibrium set up. The presence of water in the urethane will aid in certain external attacks by dilute acid and alkalies.

When carrying out chemical resistance tests, it is advisable where possible to use the same solution that the product is going to be used in. This will help cover the interpretation of the effects of several different chemicals that may be in a commercial mixture.

In the evaluation of the effect of chemicals on polyurethanes, there are two main test methods used. The first is the change in tensile properties and the second is the absorption into the material.

Changes in Volume and Dimensions

In order for samples to reach equilibrium with the test solution, the samples should not be more than 2 mm thick. The width and length should be 25 x 50 mm.

The samples are completely immersed in the liquid with no touching of the side or bottom. Laboratory glass beads will assist with this. If the test is carried out near the boiling point of the liquid, it must be carried out in a container fitted with a reflux condenser.

The change in volume can be determined gravimetrically. The change in the sample is determined by weighing in air and in a suitable liquid such as water. The volume is given by the difference between the weight in air and the weight in water. These changes can also be calculated by measurements. The thickness is determined using a micrometer and the width and length by a traveling micrometer. The results are expressed as a percentage change in volume.

The volume swell in water is normally very low. In ketones it will be high.

Changes in Physical Properties

The changes in properties are determined by preparing a number of dumbbell samples of the material under test from the same mix of polyurethane. A reference set of dumbbells is stored in the dark. The test samples are placed in a container containing at least fifteen times their volume of test liquid and kept at the desired temperature for the test period. Before immersion in the test liquid, the dimensions of the pieces must be taken.

There are two alternative test procedures.

The first requires the tensile and modulus tests to be carried out immediately after removal from the test solution and the second after drying and stabilization of the sample. The results obtained, like all tensile results, are arbitrary and used for comparison purposes. The percentage changes in properties are normally reported.

The standard hardness blocks are too thick to reach equilibrium in the normal test period, so the hardness readings are taken on a stack of 2 mm thick pieces.

The second method is applicable to situations where there is attack only from one side of the polyurethane. Typically this would be in pipe linings or thin layers of polyurethane bonded onto reinforcings. There are two basic test configurations. The first is a blind flange setup where the sample is clamped between a glass tube containing the liquid and a temperature-controlled heater. In this test two samples can be tested under identical conditions. The second is a retaining cup clamped onto the sample. Figure 10.7 illustrates the two types of test cells.

These tests can also be used to study the effect of the test liquid on the bonding of polyurethane to a reinforcing plate. In applications such as tank and pipe lining, the complete and correct bonding to the substrate is of vital importance.

The following test methods deal with the resistance to liquids:

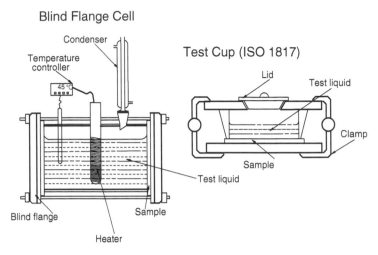

FIGURE 10.7
Fluid resistance test cells.

Method	Details
ISO 1817:2015	Rubber, vulcanized or thermoplastic – Determination of the effect of liquids
ASTM D471-12a	Standard Test Method for Rubber Property – Effect of Liquid
ASTM D1460-86(2014)	Standard Test Method for Rubber Property – Change in Length during Liquid Immersion

Heat Aging

The correlation between dry heat aging results and field aging is poor. The tests are carried out to obtain some indication of the effect of heat on the polyurethane. The sample dimensions are small compared to most applications. This causes oxidative effects to be pronounced.

In these tests the samples are placed in an oven at the set temperature (70 °C or 100 °C) for the required period of time. A control set of samples are kept in the dark at ambient. The volume of the samples in the oven should not exceed 15% of the oven's working volume. Depending on the expected results, the samples are exposed for 1, 3, 7, 10 or multiples of 7 days.

On completion of the aging tests, physical tests such as tensile, modulus at various strains, elongation at break and hardness tests may be carried out. Modern testing such as DMA tests can also be performed.

The results are often plotted as a graph against time for the properties examined.

There are a number of international tests for the determination of heat aging properties:

Method	Details
ISO 188:2011	Rubber, vulcanized or thermoplastic – Accelerated aging and heat resistance tests
ASTM D454-04(2010)	Standard Test Method for Rubber Deterioration by Heat and Air Pressure
ASTM D572-04(2010)	Standard Test Method for Rubber Deterioration by Heat and Oxygen
ASTM D573-04(2010)	Standard Test Method for Rubber Deterioration in an Air Oven
ASTM D865-11	Standard Test Method for Rubber Deterioration by Heating in Air (Test Tube Enclosure)
DIN 53508	Testing of rubber – Accelerated aging

Fungal

The Panama weathering study has shown that polyester polyurethanes are prone to fungus attack within six months. Polyether-based polyurethanes are resistant to such attacks. If specific work is required, it is best that it be carried out by a suitably competent consulting laboratory.

10.4.13   Wear Resistance

Abrasion

Abrasion tests indicate the relative resistance of the polyurethane to two-bodied wear. When choosing a test, the test most suited to the application should be used. The ranking of the results varies with the type of test used.

The DIN abrader is used extensively for wear testing. The test uses a metal drum (150 mm diameter) that is covered with a 60 mesh corundum abrasive cloth. The drum revolves at forty revolutions per minute and the sample moves down the length of the drum. The 16 mm diameter sample travels a distance of 40 m during the test. From the loss in weight of the sample and its density, the volume loss is calculated. Three runs are normally carried out to obtain an average result.

The Taber abrader is another popular tester. It is used in a variety of different fields such as the paint, textile and metal industries. A weighed sample is clamped onto a revolving table and abrasive wheels are placed onto the sample. The load on the free rotating wheels can be varied. The loss after a specified number of revolutions is recorded. The type and load on the wheels must also be stated with the result.

Many of the standards for the tests are based on commercial units:

Method	Details
ISO 4649:2010	Rubber, vulcanized or thermoplastic – Determination of abrasion resistance using a rotating cylindrical drum device
ISO 5470-1:1999	Rubber- or plastics-coated fabrics – Determination of abrasion resistance – Part 1: Taber abrader
ASTM D1630-06(2012)	Standard Test Method for Rubber Property – Abrasion Resistance (Footwear Abrader)
ASTM D2228-04(2009)	Standard Test Method for Rubber Property – Relative Abrasion Resistance by Pico Abrader Method
ASTM D3389-10	Standard Test Method for Coated Fabrics Abrasion Resistance (Rotary Platform Abrader)

Erosive Wear

Evaluation of the erosive wear of elastomers has not been standardized and most industries test the erosive wear using methods designed to simulate conditions relevant to their operations.

Organizations involved in the evaluations of erosive wear in pipes will often have a replaceable section of pipe that can have the experimental lining in it. Alternatively there may be a side loop where the experimental material can be placed.

The use of pumps running in a process stream under good supervision and instrumentation can give good results. The ideal situation is two pumps pumping the same slurry, running in parallel.

Work under I.M. Hutchings and J.C. Arnold [1],[2],[11] developed the slurry jet erosion tester. In this style of tester, a stream of erodent impinges on the surface of the sample at a predetermined angle. The mechanisms of erosive wear can be established at various angles of impingement. Quantitative work needs very careful post cleaning of the sample and the moisture equilibrium of the sample needs to be the same before and after the test.

Hector McI Clark and associates[4] [5] developed a wear tester based on a high speed disk and on a slurry pot to evaluate the nature of erosive wear. Another type of slurry pot test was described by the US Bureau of Mines [3] where only a very low angle of incidence was observed.

Walker and Bodkin [10] discuss the advantages and limitations of a number of commonly used erosive wear testers. The influence of particle size and shape is very important as well as the impingement angle and concentration of the slurry. They found that the wear rate increases with the jet velocity to the power of 2.2 (Mens and de Gee give 2.8-3.2 [7]). Wear rate is at a maximum at 30°impingement angle. The mechanism is mainly cutting. The rate increases with the size of the particle.

There are a number of other erosive wear testers used in industry to suit the type of product being evaluated. All the testers must be compared to actual field results. A major problem with a number of methods is the degradation of the slurry under test. The erosive wear of polyurethanes is very good so extended testing often needs to be carried out.

10.4.14   Dynamic Heat Buildup

Dynamic heat buildup in applications such as wheels is important. Polyurethanes with a high resilience or a low tan δ in the operating range are important. Polyurethane elastomers used in wheel applications can be evaluated using a test rig where the urethane can be run under load for a fixed period or until failure.

A typical test speed is 32 kph at a load of 55 kg. The surface temperature of the wheel can be measured using an infrared thermometer. Failure will be due to hysteresis work, and the part will decompose from the inside out.

10.5   Prototype and Service Tests

The amount of testing and verification carried out depends largely on the importance of the part being produced and its future function. An end stop for a tube may only need the appearance, hardness and the fact that it will correctly fit the tube checked. A flexible mounting of similar size will also need extensive testing to prove that it has the correct dynamic properties and can take the load, vibrations and not fail due to excessive heat buildup.

10.5.1   First Part Evaluation

The first part produced must be checked for a number of basic points:

- Correct polyurethane system used
- Molding of correct color
- Molding free of defects
- All dimensions in accordance with drawing/sketch
- Part fits correctly into rest of system

Any minor adjustments to the mold need to be made at this time; for example, the dimensions or any air pockets being removed by adding suitable sprues and/or bleed holes.

10.5.2   Verification of Design

The basic design needs to be confirmed by the customer. Does the proposed part actually look like what is required and will it fit into the rest of the equipment if it is part of a larger assembly? If not previously discussed with the supplier, any potential trouble points like sharp corners or butt joints should be brought to the attention of the customer and the potential problems resolved.

If the part is a noncritical item or if there is knowledge available from previous similar items, the part can be evaluated by the customer.

Flexibility, deformation under load and correct functioning in an assembly can be verified using the part.

10.5.3   Simulated Tests

Standard laboratory dynamic tests provide screening of grades only. They do not verify how the full-scale part will perform. Some of these tests have been developed from a generalized product situation.

The aim of the tests should be to verify the correct functioning of the part at normal and at the limits of the operating conditions. By carrying out the tests under laboratory conditions, the potential for problems in the real world is reduced. In critical applications the potential for major problems must be eliminated.

There is a large variety of service-style testing that can be carried out. These may range from very simple rigs to complicated test setups.

A roller may be subjected to pressure in a hydraulic press to simulate actual service conditions. The deflection at various loads can be noted. Wheels can be run on a test rig at service speeds under a load and the life and heat buildup recorded.

Antivibration mounts can be tested by measuring the reduction in the forces through the part under simulated working conditions. Any heat buildup must also be monitored. The tests may be carried out in an environmental chamber if nonambient temperatures are being used.

The properties of large pads can be tested in hydraulic presses such as used to cure elastomers. Any shear factors can be added using side rams.

Seals can be tested under all potential conditions of pressure, speed and environmental conditions.

The major problem in any simulated test is that of shortening the life span of the part. If a part is expected to last several years, an increase in operating temperature may introduce aging problems other than those experienced by operating at standard temperature. Similar considerations need to be applied to other variable factors such as load, vibrations or concentration of chemicals.

10.5.4   Field Trials

In certain cases more information may be obtained from field trials than by laboratory testing. Certain important criteria must always be observed before field trials commence:

- The potential for success must be very high.
- Failure of the test must not adversely affect the organization or person carrying out the test.
- The tests must be statistically designed to prevent bias.
- The length of the test must be such that an outcome can be reached.

The time span taken for a field trial to reach conclusion is important. If a part takes considerable time to wear out, intermediate evaluations must be carried out to estimate the overall life. Improvements in short-life parts are easer to evaluate.

In certain cases, field trials can provide information that testing cannot. The feel to a golfer of a polyurethane cover on a golf ball is very important to the acceptance of the new cover stock.

Field trials on mining sites and large industrial complexes have several problems of their own. The part will be required to last a long time so continued knowledge of the item must be known. A trial part on service for 12 or more months may be changed by the night shift and scrapped before being assessed. Another major problem is to ensure that the trial part and its control both see the same feed and that the operating times and conditions are rigorously recorded.

10.6   Investigative Research

There are a number of instrumental techniques that can be used to elucidate the structure of polyurethanes. These are often expensive and need specialist operation and interpretation of results.

10.6.1   Infrared Studies

The development of Fourier transform techniques with infrared spectra has made the use of infrared spectroscopy available to many more polymer chemists.

There are three main infrared bands, namely the near, mid-, and far infrared zones. The most commonly used band is the mid-range band covering 50 to 2.5 μm or in wavenumbers 200 to 4000 cm^{-1}. The near infrared that covers the range of 0.8 to 2.5 μm (wavenumbers 12500 cm^{-1} to near 4000 cm^{-1}) has overtones that have some importance in polyurethane processing.

In this region, real-time studies have been carried out to follow the course of a reaction. (wavenumbers (in cm^{-1}) = 10000 / wavelength (in μm).) The far infrared range does not have information that is relevant to polyurethane studies.

There are two ways of viewing the spectrum produced, either in the percent transmitted or in the absorption mode. For work involving the formation of urethane and urea groups, the absorption mode appears to be easier.

In the mid-range, the major technical problem is the fact that glass absorbs the infrared strongly so that special crystals such as potassium bromide, dried sodium chloride or silver chloride need to be used. In the near infrared, there are special glasses that can be used, e.g., special black glass.

Some important wavenumbers in polyurethane work follows:

Functional Group	Wavelength μm	Wavenumber cm^{-1}
-NH	3.0	3333
-NCO	4.4	2273
-C=O in esters	5.71–5.83	1751–1715
-C=O in allophanate	5.71–5.85	1751–1709
and urea	8.85–6.06	709–1650
-C=O in urethane	5.75	1739
-C=O in urea	5.9–6.1	1694–1639
Amide II band (NH)	6.5	1538
C-O in ethers	9.3	1075

Infrared spectroscopy studies the effect of infrared radiation on the molecules. In the mid-range it is the fundamental vibrations of atoms and groups of atoms. Every group has its own specific wavenumber. At the high frequency end (of the mid-range) there are hydrogen stretching vibrations (from O-H, N-H, C-H and others). As the frequencies become lower, the double bond starts absorbing (C=O and the N=C=O). At still lower frequencies there are bands due to hydrogen bonding [8].

The most commonly used technique for obtaining a spectrum is the Attenuated Total Reflectance (ATR) method in the Multiple Internal Reflectance (MIR) mode. In this method the infrared beam is passed into a special crystal of a selenide (KRS-5). The angle of incidence is such that the beam will bounce along the crystal. A sample of the polyurethane is placed hard up against the crystal. The infrared just penetrates into the material before it continues down the crystal. A number of internal bounces are obtained along the crystal. Up to 25 reflectances are obtained from a 2-mm crystal. Figure 10.8 illustrates the infrared path in an ATR cell.

Textbooks dealing with infrared will have tables with the wavenumbers for various groups. There is often confusion between "-OH" bands near the NH band at 3333 cm^{-1} and a band from atmospheric contamination near the NCO at 2273 cm^{-1}.

It must always be noted that the wavenumbers will vary from instrument

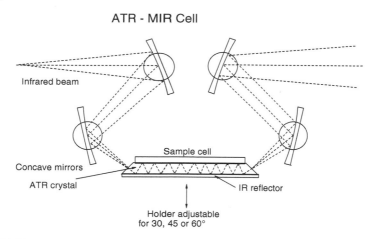

FIGURE 10.8
Infrared light path in an ATR cell.

to instrument. There are also influences from other functional groups attached to the group that will change the exact position.

A qualitative use for infrared spectra is the identification of the type of system used based on standard databases such as the Sadler library of reference spectra or by comparing with a database developed internally. A large library of spectra taken on the same instrument and under the same conditions will give the best results.

The type of TDI used can be determined by studying the bands at 810 cm^{-1} and 782 cm^{-1}. 100% 2,4-TDI has a peak at 810 cm^{-1} and no peak at 782 cm^{-1}. As the percentage 2,6-isomer increases, the band at 782 cm^{-1} intensifies. The ratio of these two bands indicates the grade of TDI used.

The reaction rates of systems can be measured. Using heated cells the reaction between mixtures of polyols and isocyanates can be followed. There are a number of changes to the infrared spectrum that takes place. The -OH band at 3460 cm^{-1} will decrease, as will the -NCO band at 2270 cm^{-1} as the diisocyanate reacts with the polyol, leaving only the terminal isocyanate groups. Other urethane bands at 3289, 1730, 1534 and 1230 cm^{-1} will increase. The ether group (1112 cm^{-1}) will stay the same because it does not take part in the reaction. It can be used as a reference to normalize the other bands. If a polyol and MDI are studied, it is found that there is only one rate constant, whereas when using 80:20 TDI, there are two rate constants.

Similar work can be carried out during the curing by studying the reaction rate between the curative and the prepolymer. The "-NCO" band at 2270 cm^{-1} will decrease and the urethane or the urea bands will increase, depending on the type of curative used. This also shows the development of hydrogen bonding in the system. The urethane and urea bands will split into bonded and nonbonded bands. Work by Seymour and Estes has shown how the percentage hydrogen bonding can be calculated [9].

The effect of temperature on hydrogen bonding can also be examined. It has been found that the bonding disappears on heating and returns as the material cools. With moderate heating, the level of hydrogen bonding will increase when the material cools down.

There are glasses that are transparent to infrared in the near-infrared range. The special glass is made into a fiber-optic cable enabling some flexibility. Overtones from the mid-infrared range can be used to monitor reactions. A disposable probe can be inserted in or at the surface of a polyurethane molding. This technique is particularly suited to very fast cures.

10.6.2 Gas Chromatography

A withdrawn ASTM standard method D3432-89(1996)e1 describes how to determine the amount of unreacted TDI in prepolymers and coating solutions. An internal standard is used to normalize the results. Care must be taken not to get any prepolymer on the main column.

This test is important as the unreacted isocyanate is a prime cause of health problems.

10.6.3 Nuclear Magnetic Resonance (NMR)

This method relies on the fact that the isotopes of certain atoms have electrons that will flip under certain conditions. This change in state can be detected and how they are connected can be shown. The ^{1}H and ^{13}C isotopes are able to produce resonance spectra for polyurethane raw materials and cured samples. There is a standard to determine the primary hydroxyl content of polyols:

Method	Details
ASTM D4273-11	Standard Test Method for Polyurethane Raw Materials: Determination of Primary Hydroxyl Content of Polyether Polyols

Any fluid samples can be tested on standard NMR machines. The structure of the polyol can be calculated and any branching will be shown. The approximate size and structure of prepolymers can also be determined using both ^{1}H and ^{13}C NMR methods.

Solid polyurethanes can be tested using specialized NMR instruments and the overall structure determined.

10.6.4 X-Ray Diffraction

Various x-ray diffraction techniques have been used to study the size and distribution of hard and soft segments near the surface of the polyurethane.

10.6.5   Differential Scanning Calorimetry (DSC)

The amount of heat absorbed or given out by a sample of polyurethane is measured as the sample is heated over a range of temperatures. Changes in the state of the polyurethane occur at various temperatures. Changes occur when the hard, then soft segment becomes mobile and when hydrogen bonding starts to break down.

Samples are normally heated to just above the softening point and then allowed to cool down before a test is carried out. With castable polyurethanes, decomposition takes place when the sample is softened so the annealing cycle must either not be used or the temperatures not taken up as high.

10.6.6   Atomic Force Microscopy

Atomic force microscopy has been used to evaluate the nature and size of the polyurethane structure at the surface of a sample.

In this method the surface of the sample is coated with a layer of gold and a probe scans over the surface. As the nature of the chemistry below the probe changes, the force field acting on the probe changes. The hard segment clusters can be visualized.

10.6.7   Scanning Electron Microscopy (SEM)

Scanning electron microscopy is an important tool when examining the mode of wear of any sample. The surface of the sample is coated with a very thin layer (only several atoms thick) of a conductive material such as gold. The surface is then scanned using a beam of electrons and the image magnified and recorded.

SEM gives results with a much greater depth of field than conventional optical microscopy. When wear surfaces are being examined, a full-depth image of the wear damage can be seen. The representation gives a very three-dimensional effect allowing for very good visualization.

10.6.8   High Performance Liquid Chromatography (HPLC)

HPLC instrumentation and methods have taken over from many GLC methods and using the correct method give better resolution and accuracy. HPLC methods are used in the urine analysis of people who have been exposed to MOCA.

HPLC techniques have been developed to analyze for free TDI in prepolymers as an alternative to the previous ASTM GLC method.

If the HPLC is set up for the determinations of anions and cations, the metallic catalyst level in quasiprepolymers can be determined.

10.6.9 Size Exclusion Chromatography

In this variation of high performance liquid chromatography (HPLC), the columns are packed with material that will hold back molecules dependent on the molecular size. Using the correct solvent system and column types, the molecular weight and distribution of the polyols can be determined. Standard samples are needed to calibrate the system. The molecular weight distribution of prepolymers can also be determined. This will enable an estimation of the number of soft segment chains there are.

References

[1] J. C. Arnold and I. M. Hutchings. The mechanisms of erosion of unfilled elastomers by solid particle impact. *Wear*, 138(1-2):33–46, June 1990.

[2] J. C. Arnold and I. M. Hutchings. A model for the erosive wear of rubber at oblique impact angles. *J. Phys.D;Appl.Phy*, 25(1A):A222–A229, 1992.

[3] R. Blickensderfer and J. H. Tylczak. Lab wear testing capabilities of the Bureau of Mines. *Bureau of Mines circular 9001*, IC9001:1–41, 1987.

[4] H. M. Clark and J. Tuzsion. Measurements of specific elements for erosive wear using a coriolis erosion tester. *Wear*, 241:1–9, 2000.

[5] H. M. Clark and K. K. Wong. Impact angle, particle energy and mass loss in erosion by dilute slurries. *Wear*, 186-187(2):454–464, 1995.

[6] D. W. Cumberland. Panama weathering study by DuPont Corporation. *Trade Literature*, pages 1–22, 1985.

[7] J. W. M. Mens and A. W. J. De Gee. Erosion in sea water sand slurries. *Tribology International*, 19(2):59–61, 1986.

[8] G. Pruckmayr. FTIR in polyurethane technology. In *PMA*, pages 1–23, Portland Oregon., 1993. Polyurethane Manufacturers Asscociation.

[9] R. W. Seymour, G. M. Estes, and S. L. Cooper. Infrared studies of segmented polyurethane elastomers. I. Hydrogen bonding. *Macromolecules*, 3(5):579–583, September 1970.

[10] C. I. Walker and G. C. Bodkin. Erosive wear characteristics of various materials. In *12th International Conference on Slurry Handling and Pipeline Transport*, pages 191–210, Brugge, Belguim, 1993.

[11] J. B. Zu, I. M. Hutchings, and G. T. Burstein. Design of a slurry erosion test rig. *Wear*, 140(2):331–344, November 1990.

Part V

Appendices

Appendix A

Abbreviations and Trade Names

Name	Description
Acclaim Polyol	Low monal PPG
BDO	1,4-Butane diol
Benzoflex 9-88 SG	Dipropylene glycol dibenzoate
CHDI	Cyclohexyl diisocyanate
Crosslink 30/10	HQEE
DABCO	DABCO 33LV
DABCO 33LV	1,4-Diazabicyclo[2.2.2]octane/ triethylene diamine
DIOP (DOP)	Diisooctal phthalate
Ethacure 300	3,5-Dimethyl-2,4-toluenediamine and 3,5-Dimethyl-2,6-toluenediamine
Fyrol PCF	Tri(b-chloroisopropyl) phosphate
HDI	Hexamethlyene diisocyanate
HQEE	Hydroquinone bis-(beta hydroxyethyl)ether
Isonol 93	Polyether triol — now Conap AH50
Lonzacure M-CDEA	4,4'-Methylenebis(3-chloro-2,6-dichloroaniline)
MBOCA	3,3'-Dichloro-4,4'-diaminodiphenylmethane
MDI	4,4'-Diphenylmethane diisocyanate
Messamol oil	Alkyl sulfonic acid ester of phenol (Miles)
MOCA	3,3'-Dichloro-4,4' diaminodiphenylmethane
m-Pryol	2-Methyl-2-pyrrolidone (CAS 872-50-04)
NDI	1,5-Naphthalene diisocyanate
NMP	2Methyl-2-pyrrolidone (CAS 872-50-04)
Polacure	Trimethylene glycol di-p-aminobenzoate
PPDI	Para-phenylene diisocyanate
PPG	Polypropylene ether glycol
PTMEG	Polytetramethylene ether glycol
Santicizer 160	Butyl benzyl phthalate
Stabaxol	Anti hydrolysis agent Rhein Chemie
TCP	Tricresyl phosphate (CAS 1330-78-5)
TDI	2,4-Toluene diisocyanate 2,6-Toluene diisocyanate
Terathane	Polytetramethylene ether glycol
THF	Tetrahydrofuran
TMP	Trimethylol propane
TODI	3,3'-Dimethyldiphenyl 4,4'-diisocyanate

Appendix B

Polyurethane Curatives

Curative	Equivalent Weight	Comment
1,4-Butane diol	45.1	BDO
AH40	133.6	Commercial formula
Baytec 1604	121.4	
BDO	45	1,4 Butane diol
Caytur 21	653	MDA Complex
Conap AH40	133.5	Commercial formula
Conap AH50	90	Previously known as Isonol 93
Conap AH33	280	
Curene 3005	280	
Cyanacure	138.2	1,2-bis[2-amino-phenylthio]ethane
DEG	53	Diethylene glycol
ETDA	89.2	Diethyltoluene diamine
Ethacure E100	89.2	DETDA
Ethacure E300	107.2	Di-(methylthio)toluene diamine DMDTA
Ethylene glycol	31.03	
Glycerol	30.1	Glycerine
HER	99	
HQEE	99.1	Hydroquinone bis-(beta-hydroxy-ethyl)ether
Isonol 93	90	Range 88 − 93 triol
M-CDEA	189.7	Lonzacure
MDA	99.1	Methylene dianiline
MMEA	141.2	
MOCA (MBOCA)	133.6	
MOEA	127.2	
Plurocol TP-440	145	Range 139 − 145 (triol)
Polacure 740	157.2	
Poly-cure 1000	155	
PPG 1000	500	Polypropylene glycol
PTMEG 1000	500	
t-BTDA	89.4	

Curative	Equivalent Weight	Comment
TIPA	64	Triisopropane amine
TMP	44.7	Trimethyol propane
Unilink 4200		Commercial R groups
Versalink 740	157	
Versalink P-1000	60 0	
Versalink P650	415	

Appendix C

Mold Release Agents

Name	Manufacturer	Comments
200® and 20	Dow Corning®	
3422 Polyurethane release	Dow Corning	Water-based
Cilrelease 910	Chemical Innovation	Concentrated mold at 50°C
Crystal 4000	TSE Industries	
Eralease	Era Polymers	
Freecote #1 NC	Loctite	Teflon-based
Freecote Aqualine® PUR 100	Loctite	
Freecote FRP-NC	Loctite	
Freecote® WOLO	Loctite	
Gorapur®	Degussa	
Mavcoat® mold release	Marerix Solutions Inc.	
Mold release 81801	Chesterton	
Mylar film	DuPont	
Nonstikenstoffe	Freekote	
Rhodosil 47V50	Rhodia	

® 200 Fluid and Dow Corning registered trademark of Dow Corning Corporation

® Aqualine registered trade name of Henkel Loctite Corporation, Düsseldorf, Germany

® Frecote. registerd trade name of Henkel Loctite Corporation, Düsseldorf, Germany

® Gorapur registered trade name of Degussa

Appendix D

Calculations

D.1 Molecular Weight

The mole (mol) is the fundamental unit describing the amount of a chemical species. The weight of one mole of a substance is its gram formula weight (fw), which is the summation of the atomic weight in grams for all the atoms in the chemical formula of the species.

Molecular weight	=	C_7H_8
Formula weight	=	92.14
Nominal Mass	=	92

D.2 Equivalent Weight

The equivalent weight is based on the behavior of the substance in any particular reaction. The equivalent weight of a substance is the result of the formula weight (molecular weight) divided by the number of reactive sites in it taking part in the reaction.

The equivalent weight of a diol with a molecular weight of 1500 will be $1500/2 = 750$.
If it is a triol with three reactive sites, the equivalent weight will be $1500/3 = 500$.

Specification sheets will often give the equivalent weight of the polyol in terms of the hydroxyl number ("OH" value). The hydroxyl number is defined as the number of milligrams of potassium hydroxide (KOH) equivalent to the hydroxyl content of 1.0 grams of polyol. Manipulation of this definition gives the following equation:

$$\text{Equivalent weight of any polyol} = (56.1)(1000)/\text{OH number}$$

where
  56.1 g = equivalent weight of KOH
  1000 = mg / 1.00g

If the OH value of a polyol is given as 28.05, the equivalent value will be

$$\text{Equivalent weight of the polyol} = (56.1)(1000)/28.05$$
$$= 2000$$

If the molecular weight is known to be 4000, the number of functionality or reactive sites will be

$$\text{Functionality} = \text{Molecular weight} / \text{Equivalent weight}$$
$$= 4000/2000$$
$$= 2$$

D.3   Equivalent Weight of Blends of Polyols or Curatives

If a blend of items is used, the equivalent weight of each item must be taken into account when calculating the final value:

$$\text{EW of blend} = \frac{\text{Wt. of A} + \text{Wt. of B}}{\dfrac{\text{Wt. of A}}{\text{Ew. of A}} \dfrac{\text{Wt. of B}}{\text{EW of B}}}$$

For example, if a mixture of 3 parts TMP is used with 1 part TIPA, then

$$\text{EW of blend} = \frac{3 + 1}{\dfrac{3}{44.7} + \dfrac{1}{63.7}}$$
$$= 48.3$$

The above calculation gives a working EW of 48.3.

D.4 Weight Calculations

There are two main methods to calculate the weight of the polyol and diiso-cyanate to use in a reaction:

- Molecular ratios

- To fixed NCO level

Molecular Ratios

In this method the number of gram moles of each item is taken into account when making the first approximation of the weight ratios.

For example, if two moles of TDI were to react with one mole of 1000 MW PPG:

$$EW = 87.1 + 500 + 87.1$$
$$\text{Total wt } EW = 87.1 + 500 + 87.1$$
$$= 174.2 + 500$$
$$= 674.1$$

If it is desired to make 2.5 kg of the prepolymer, the following portions would have to be reacted together:

$$\text{Weight mixture} = 2.5 \times 1000 = 2500\text{g}$$
$$TDI = (174.2/674.2) \times 2500 = 646\text{g}$$
$$PPG = (500/674.2) \times 2500 = 1854\text{g}$$

It must be remembered that these are theoretical calculations and final results may be different due to raw material purities, transferring efficiencies and the degree of conversion.

Fixed NCO Levels

To produce a prepolymer with 0%, free isocyanate (NCO), one equivalent of the polyol with one equivalent of the isocyanate must be reacted together:

$$\%\text{NCO of prepolymer} = \frac{\text{Wt. of teminal NCO} \times 100}{\text{Total Wt.}}$$

From first principles:

The equivalent weight of NCO = 14 + 12 + 16 = 42 (more exactly, 42.02)

The equivalent weight of TDI = 87

The equivalent weight of MDI = 125

$$\%\text{NCO in an isocyanate} = \frac{42 \times 100}{\text{Equivalent wt. isocyanate}}$$

$$\%\text{NCO in an TDI} = \frac{42 \times 100}{87} = 48.3(Z)$$

$$\%\text{NCO in an MDI} = \frac{42 \times 100}{125} = 33.6(Z)$$

For example,
It is desired to make a prepolymer with a final NCO of 20% from MDI and a 500 MW polyol:

$$\text{Diol (D)} = 500 \text{ g EW}$$
$$\text{MDI (I)} = 125 \text{ g EW}$$
$$\text{Total g EW (T)} = 625 \text{ g EW}$$

To produce a 0%, one must react one equivalent of polyol with one equivalent of isocyanate:

$$\text{Total weight (T)} = D + I$$

If a final NCO (N) level of 20% is desired, additional isocyanate (Y g EW) must be added:

$$N = \frac{Z \times Y}{T + Y}$$

$$20 = \frac{33.6 \times Y}{635 + Y}$$

$$20(625) + 20Y = 33.6Y$$

Solving for Y:

$$33.6Y - 20Y = 20(625)$$
$$13.6Y = 12500$$
$$\text{Addition MDI (Y)} = 919\text{g}$$
$$\text{The total MDI} = 125 + 919 = 1044\text{g}$$
$$\text{The total Polyol} = 500\text{g}$$

These calculations provide a starting point. Final adjustments need to be made depending on raw material and processing conditions.

D.5 Acid Levels

To prevent gelation and side reactions it is normally recommended that the reaction be kept very slightly acid. The level is 0.33 microequivalents per gram of prepolymer.

To convert to microequivalents:

$$\text{EW HCl} = 36.5$$
$$\text{EW KOH} = 56.1$$

$$\text{Isocyanate acid} = \frac{\text{Specified ppm} \times \text{wt. diisocyanate (in grams)}}{\text{g EW HCl}}$$

The acidity of the diisocyanates is normally specified in terms of parts per million of isocyanate acid in the diisocyanate. Documentation will specify the acidity either in ppm or as a percentage:

$$\text{ppm} = \frac{X\% \times 10^6}{10^2}$$

Thus,

$$0.004\% HCl = 0.004 \times 10^4 = 40\text{ppm}$$

Using the previous example for 20% total NCO:
MDI Weight = 1044 g Acidity 5 ppm as HCl
Diol 500 EW Weight = 500 g Acid value 0.0015 mg KOH/g

$$\text{Acidity of MDI} = \frac{5 \times 1044}{36.5} = 143.0 \text{ microequivalents of acid}$$

$$\text{Polyol base level ppm} = \frac{0.015 \times 10^6}{1000} = 15 \text{ ppm}$$

$$\text{Polyol base level} = \frac{15 \times 500}{56.1} = 133.7 \text{ microequivalents of base}$$

To have the desired level of 33 microequivalents/g there must be 0.33 x 1544 = 510 microequivalents of acid. Without any additions there will be a very slight excess of acid but less than the desired level of 0.33 per gram. Therefore more acid must be added, using benzoyl chloride (EW 140.6) as the acid.

$$\text{Benzoyl chloride} = ((510 - (1430 - 134) \times 140.6)/1000 = 70.4 \text{ mg}$$

If 85% phosphoric acid is used instead, the formula needs to be adjusted.

Appendix E

Isocyanate Calculation

Determination of %NCO in Prepolymer

The method for the determination of the isocyanate level is based on that given by Wright and Cumming.

Background

The NCO content of diisocyanates can be determined by the reaction with an excess of a standard solution of dibutylamine in chlorobenzene (or dry toluene) and titration of the excess with hydrochloric acid.

Reagent preparation

The dibutylamine is prepared by dissolving 129 g of freshly distilled dibutylamine in 871 g of redistilled chlorobenzene and storing the resultant solution in the dark.

The solution is relatively stable but it is advisable to determine the amine content weekly by titration with aqueous 1 N hydrochloric acid in methanol.

1M Hydrochloric acid – standard laboratory-grade 1 molar hydrochloric acid is suitable (N).

Bromophenol Blue indicator – use 1% alcoholic solution.

Method

Weigh 2 grams of sample into a conical flask. Record actual weight to 0.001 grams (W).

Add sufficient dry solvent (chlorobenzene or toluene) (approximately 5 ml) to dissolve sample. Warm gently and agitate if required.

Add 25.0 ml of the standard dibutylamine to the dissolved sample with a pipette. The reaction is rapid and takes only a few minutes for completion when a clear solution is obtained.

After adding two or three drops of Bromophenol Blue and 100 ml of methanol, titrate the excess of dibutylamine with 1N hydrochloric acid (V1).

Carry out blank titration without the sample (V2).

$$\%\text{NCO} = 4.20 \times N \times \frac{V2 - V1}{1000 \times W} \times 100$$

Comments The terms 1 M and 1 N refer to the same concentration of hydrochloric acid in solution. 1 M is one gram molar of hydrochloric acid (HCl) in 1 liter. N is an abbreviation for normality based on the number of active hydrogens in the compound. HCl has one, therefore normal is numerically equal to molar.

Appendix F

Chemical Structures

Aromatic Isocyanates
2,4 TDI

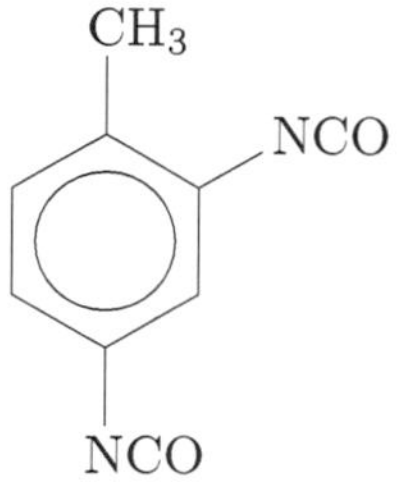

Name	TDI
Chemical name	2,4-Diisocyanato-1-methylbenzene
CAS No.	584-84-9 121-14-2
EINECS	204-450-0
Formula Wt.	174.2

2,6 TDI

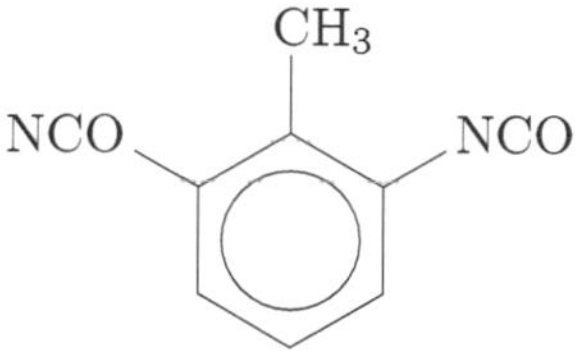

Name	2,6 TDI
Chemical name	1,3-Diisocyanato-2-methylbenzene
CAS No.	91-08-7
EINECS	202-039-0
Formula Wt.	174.2

4-4′ MDI (MDI)

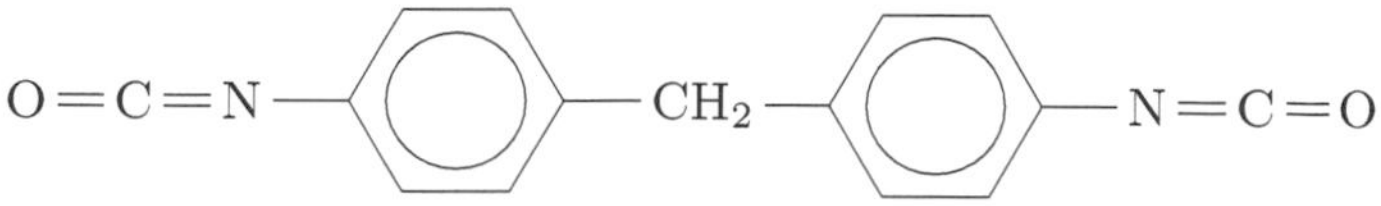

Name	MDI
Chemical name	1,1′ Methylene bi-(4-isocyanatobenzene)
CAS No.	101-68-8
EINECS	202-966-0
Formula Wt.	250.3

2,4′ MDI

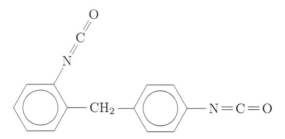

Name	2,4 MDI
Chemical name	1-isocyanato-2-(4isocyanatobenzyl)benzene
CAS No.	5873-54-1
EINECS	227-534-9
Formula Wt.	250.3

PPDI

Name	PPDI
Chemical name	1,4-Diisocyanayobenzene
CAS No.	104-49-4
EINECS	203-207-6
Formula Wt.	160.1

NDI

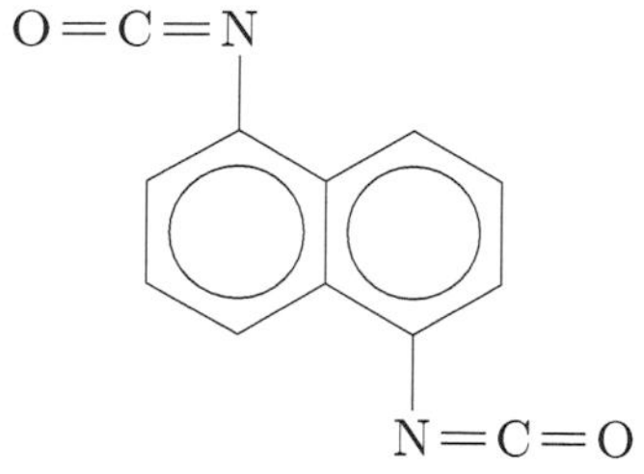

$$O=C=N$$

$$N=C=O$$

Name	NDI
Chemical name	1,5-Diisocyanatonaphthalene
CAS No.	3173-72-6
EINECS	173-72-6
Formula Wt.	210.2

TODI

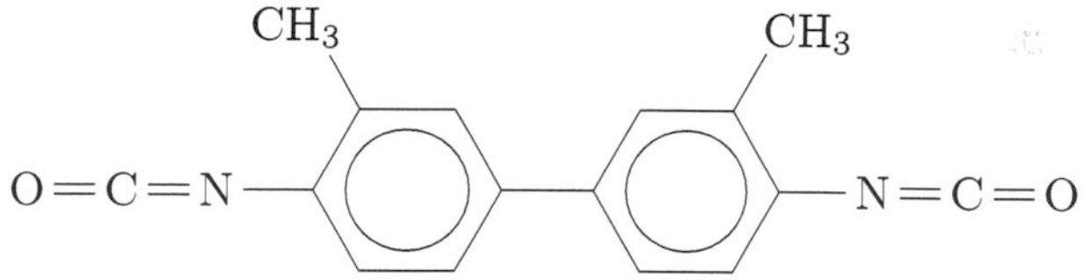

$$CH_3 \qquad CH_3$$
$$O=C=N- \qquad -N=C=O$$

Name	TODI
Chemical name	4,4'-Diisocyanto-3,3'-dimethylbiphenyl
CAS No.	91-97-4
EINECS	202-112-7
Formula Wt.	264.3

Aliphatic Isocyanates
H12MDI

$$O=C=N-\text{(cyclohexane ring)}-CH_2-\text{(cyclohexane ring)}-N=C=O$$

Name	H12MDI
Chemical name	1,1′-Methylenebis(4-isocyanatocyclohexane)
CAS No.	5124-30-1
EINECS	225-863-2
Formula Wt.	262.3

CHDI

$$O=C=N-\text{(cyclohexane ring)}-N=C=O$$

Name	CHDI (Hylene CHDI)
Chemical name	1,4-Diisocyanatocyclohexane
CAS No.	—
EINECS	—
Formula Wt.	166.2

HDI

$$O=C=N-CH_2-CH_2-CH_2-CH_2-CH_2-CH_2-N=C=O$$

Name	HDI
Chemical name	1,6-Diisocyanatohexane
CAS No.	822-06-0
EINECS	931-274-8
Formula Wt.	168.2

IPDI

$$N=C=O$$ (on cyclohexane ring)

$$CH_3,\ CH_3 \quad\quad CH_2-N=C=O,\ CH_3$$

Name	IPDI
Chemical name	5-Isocyanato-1-(isocyanatomethyl)-1,3,3-trimethylcyclohexa
CAS No.	4098-71-9
EINECS	223-861-6
Formula Wt.	222.3

H$_6$XDI

$$N=C=O$$

$$N=C=O$$

H$_6$XDI

Name	H$_6$XDI
Chemical name	1,3-diisocyanatocyclohexane
CAS No.	38661-72-2
EINECS	—
Formula Wt.	166.2

Amine curatives
Caytur 21 DA

$$3 \left[NH_2 - \bigcirc - CH_2 - \bigcirc - NH_2 \right] NaCl$$

Name	Caytur 21 DA
Chemical name	3:1 complex of MDA and NaCl in DOA
CAS No.	21646-20-8
EINECS	244-493-2
Equivalent Wt.	178-185

Caytur 31 DA

Name	Caytur 31 DA
Chemical name	47% MDA/NaCl complex in DOP
CAS No.	Trade secret
EINECS	—
Equivalent Wt.	250
Warnings	Class 2 Carcinogen

MOCA, MBOCA

$$H_2N - \bigcirc\!\!\!\overset{Cl}{} - CH_2 - \bigcirc\!\!\!\overset{Cl}{} - NH_2$$

Name	MOCA (MBOCA)
Chemical name	4,4'-Methylenebis(2-chloroaniline)
CAS No.	101-14-4
EINECS	202-918-9
Equivalent Wt.	133.6
Warnings	Class 2 Carcinogen

Comments available in a number of different purities and forms.

Unilink 4200

$$R-HN-\langle\bigcirc\rangle-CH_2-\langle\bigcirc\rangle-NH-R$$

Name   Unilink 4200
Chemical name
CAS No.   5285-60-9
EINECS   403-240-8
Equivalent Wt.   107.3

Cyanacure

$$\langle\bigcirc\rangle-S-CH_2-CH_2-S-\langle\bigcirc\rangle$$
$$NH_2 \qquad\qquad NH_2$$

Name   Cyanacure
Chemical name   1,1'-(ethane-1,2-diyldisulfanediyl)bis(2-isocyanatobenzene)
CAS No.   52411-33-3
EINECS   257-901-9
Equivalent Wt.   138.2

M-CDEA

$$CH_3 \quad Cl \qquad Cl \quad CH_3$$
$$H_2N-\langle\bigcirc\rangle-CH_2-\langle\bigcirc\rangle-H_2N$$
$$CH_3 \qquad\qquad CH_3$$

Name   M-CDEA
Chemical name   4,4-Methylene-bis-(3-chloro-2,6-diethylaniline)
CAS No.   106246-33-7
EINECS   402-130-7
Equivalent Wt.   189.7

Polacure 740M

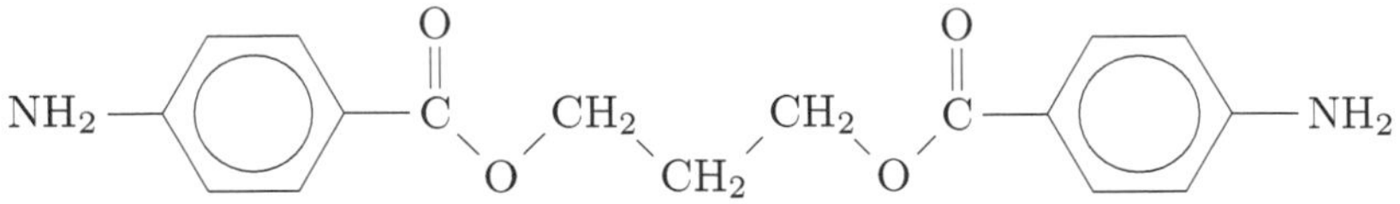

Name	Polacure 740M
Chemical name	Propane-1,3 diyl(bis(4-aminobenzoate)
CAS No.	57609-64-0
EINECS	260-847-9
Equivalent Wt.	157

Baytec 1604

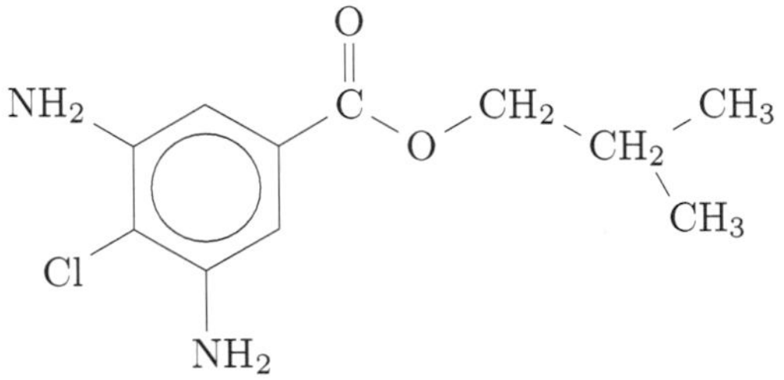

Name	Baytec 1604
Chemical name	Butyl 3,5-diamino-chlorobenzoate
CAS No.	32961-44-7
EINECS	251-311-5
Equivalent Wt.	242.7

Ethacure 100

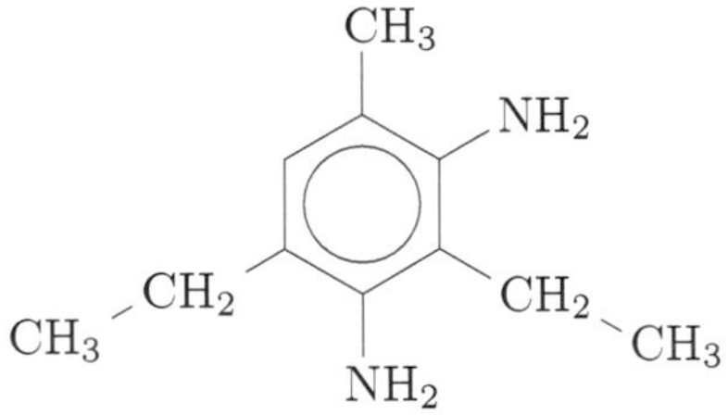

2,4 isomer ±80%

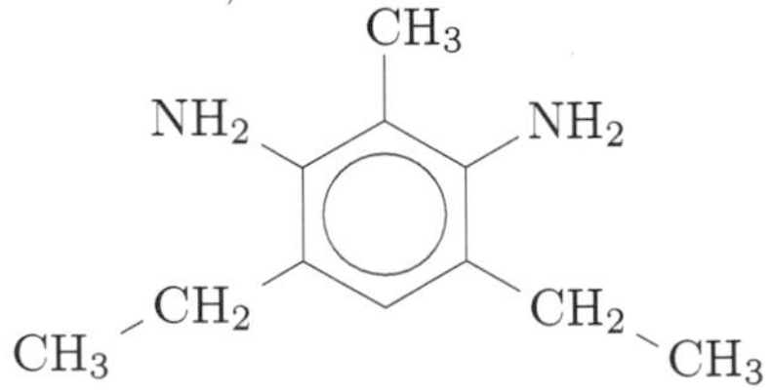

2,6 isomer ±20%

Name	Ethacure 100 (DETDA)
Chemical name	80% 2,4-diethyl-6-methylbenzene-1,3-diamine
	20% 4,6-diethyl-2-methylbenzene-1,3-diamine
CAS No.	68479-98-1
EINECS	—
Equivalent Wt.	178.3

Ethacure 300

2,4-isomer ±80%   2,6-isomer ±20%

Name	Ethacure 300 (DMTDA)
Chemical name	Dimethylthiotoluenediamine
CAS No.	106264-79-3
EINECS	403-240-8
Equivalent Wt.	107.3

Conacure AH 40

Name	Conacure AH 40
Chemical name	Commercial
CAS No.	Commercial
EINECS	—
Equivalent Wt.	133.3

Hydroxyl Curatives
Ethylene glycol, EG

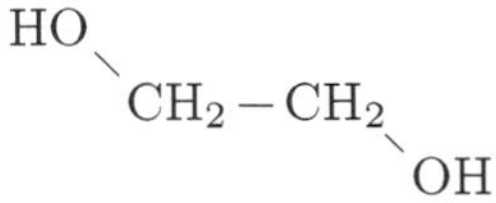

$$HO-CH_2-CH_2-OH$$

Name	Ethylene glycol
Chemical name	1,2-Ethanediol
CAS No.	107-21-1
EINECS	203-473-3
Equivalent Wt.	31.05

1,4 BDO

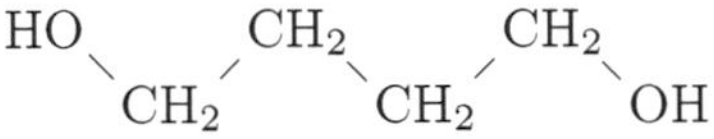

$$HO-CH_2-CH_2-CH_2-CH_2-OH$$

Name	1,4 BDO
Chemical name	1,4-Butane-diol
CAS No.	110-63-4
EINECS	203-786-5
Equivalent Wt.	45.05
Warnings	The material is hygroscopic

Glycerine

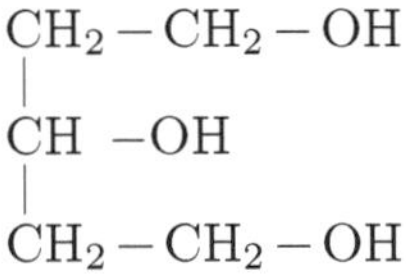

$$CH_2-CH_2-OH$$
$$CH-OH$$
$$CH_2-CH_2-OH$$

Name	Glycerine
Chemical name	Glycerol
CAS No.	56-81-5
EINECS	200-289-5
Equivalent Wt.	30.03

TMP

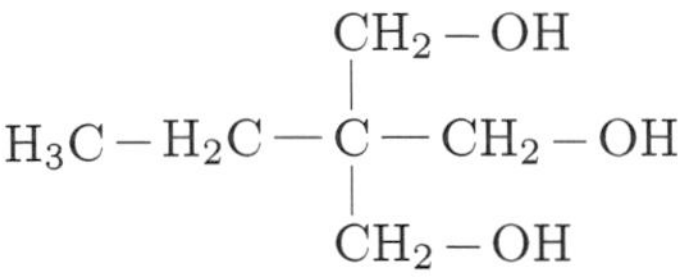

$$H_3C-H_2C-\underset{\underset{CH_2-OH}{|}}{\overset{\overset{CH_2-OH}{|}}{C}}-CH_2-OH$$

Name	TMP
Chemical name	Trimethylolpropane
CAS No.	77-99-6
EINECS	201-074-9
Equivalent Wt.	44.7

Polyol TP 30

$$CH_3 - C \begin{cases} - CH_2[-O-CH_2-CH_2]_n\text{-}OH \\ - CH_2[-O-CH_2-CH_2]_n\text{-}OH \\ - CH_2[-O-CH_2-CH_2]_n\text{-}OH \end{cases}$$

Name	Polyol TP-30
Chemical name	Ethoxylated trimethylolpropane
CAS No.	50586-59-9
EINECS	—
Equivalent Wt.	88–96

Comments: Also known as Voranol 234-630, Isonal 93 and AH50.

HQEE

$$OH-CH_2-CH_2-O-\bigcirc-O-CH_2-CH_2-OH$$

Name	HQEE
Chemical name	Hydroquinone-bis(2-hydroxyethyl)ether
CAS No.	104-38-1
EINECS	203-197-3
Equivalent Wt.	99.1

Appendix G

Applications

Industry	Application	Comments
Automotive	Bearings	
Automotive	Flexible couplings	
Automotive	Grommets	
Automotive	Suspension bushes	
Building	Concrete panel molds	
Building	Gate seals	
Building	Lift centralizing wheels	
Building	Paver molds	
Chemical	Hydrocyclones	
Domestic	Automatic gate wheels	Very hard – good compression set
Domestic	Bowling balls	
Domestic	Golf ball cleaners	
Domestic	In-line roller wheels	
Domestic	Roller skate wheels	High resilience MDI ether
Electrical	Antistatic pads	
Electrical	Potting	
Engineering	"O" rings	
Engineering	Flange seals	
Engineering	Gear wheels	
Engineering	Hydraulic seals	Diaphragms
Engineering	Lip seals	
Engineering	Location fixtures	
Engineering	Press brake pads	
Engineering	Shock absorption pads	
Engineering	Stripper plates	
Engineering	Vibration absorption pads	
Engineering	Wiper blades	
Food	Food conveyor systems	FDA wet food contact – MDI ester
Food	Grain handling equipment	FDA approval, abrasion resistance MDI ester
Food	Meat processing equipment	FDA wet food contact MDI ester

Industry	Application	Comments
Industrial	Cable guides	
Industrial	Flattening rollers	
Industrial	Fork-lift tires	Low heat buildup TDI ether
Industrial	Hammers	Tear resistance, low resilience TDI ester
Industrial	Log Pushers	
Industrial	Printing rolls	Solvent resistance at low DurometerTDI ester
Industrial	Sand blast curtains	High resilience - impingement abrasion MDI ether resistance
Industrial	Wheels and castors	
Mining	Bucket liners	
Mining	Chute linings	
Mining	Conveyor heads	
Mining	Crossover pads	
Mining	Cyclone spigots	
Mining	Cyclones	
Mining	Filter press parts	
Mining	Flotation cell liners	
Mining	Flow control valves	
Mining	Grading screens	High abrasion resistance
Mining	Idler wheels	
Mining	Launder equipment	Hydrolysis resistance MDI ether
Mining	Pipe linings	
Mining	Pump impellers	
Mining	Pump liners	
Mining	Pump throat	
Mining	Pump back plates	
Mining	Scraper knives	
Oil	Oil pipeline pig	Oil abrasion resistance TDI ester
Paper Mills	Rolls	Hydrolysis resistance, hardness stability TDI ether
Pipelines	Cleaning pigs	
Pipelines	Sealing rings	

Appendix G

Glossary

Abrasion Wear due to friction.

Activator A compounding material used in small proportions to increase the effectiveness of an accelerator.

Additive A material combined with the prepolymer and curative to modify the final properties of the casting. This includes plasticizers, fillers and stabilizers.

Adhesive failure A separation of two bonded surfaces within the bonding material.

Aliphatic A class of organic compounds containing straight, branched or cyclic fragments of carbon atoms.

Amine A class of compounds used as catalysts or curatives in polyurethane foam and elastomer reactions. Amines are characterized by having N, NH or NH_2 groups in the molecule.

Amine equivalent The same as the equivalent weight of the prepolymer and is equal to the (formula wt. NCO/ %NCO).

Annealing The process of heating and cooling metals and some plastics to reduce brittleness and/or improve strength.

Antioxidant Compounding material used to retard deterioration caused by oxidation.

Antiozonant Compounding material used to retard deterioration caused by ozone.

Aromatic A class of organic compounds containing a resonant, unsaturated ring of carbon.

Benchmarks Marks of known separation applied to a specimen and used to measure strain.

Bloom A discoloration or change in appearance of the surface of a polyurethane product caused by the migration of a liquid or solid to the surface.

275

Blowing The process of foaming polyurethane.

Blowing agent A substance incorporated in a mixture for the purpose of producing foam. Polyurethane foam can be produced using a physical blowing agent or a chemical blowing agent.

Bund Low enclosure around tanks or drums to prevent escape of liquid.

Certificate of Analysis (C of A) Documentation of analytical test results for a chemical or chemical system.

CAS Number A number assigned by the Chemical Abstract Service that is a unique identifier for each chemical.

Catalyst A chemical that initiates or speeds up a chemical reaction.

Cell A single small cavity surrounded partially or completely by walls.

Cell closed A cell totally enclosed by its walls and not interconnecting with other cells.

Cell open A cell not completely enclosed by its walls and interconnecting with other cells.

Chain extenders Low molecular weight molecules that usually react with diisocyanates. They form rigid, crystalline, hard segments in the polyurethane and lengthen the main urethane chain by end-to-end attachment.

Coefficient of friction A material property of the contacting surfaces and of the contaminants and other films at their interface.

Coefficient of thermal expansion The rate at which a material expands per degree of temperature.

Cold cure Casting process for the production of elastomers, in which the elastomer is mixed, dispensed and cured at or near room (ambient) temperature.

Compound An intimate admixture of polymer(s) with all the materials necessary for the finished article.

Compression set A measure of permanent deformation remaining in an elastomer or flexible foam after a deforming force is removed. For most applications, a low degree of compression set is desirable.

Crazing Fine cracks on surface of a material.

Creep Compressive deformation occurring over time in both cured and uncured polyurethane, resulting from the application of a constant load or stress.

Cross-linking Formation of chemical bonds or bridges between separate polymer chains, resulting in a three-dimensional polymer network.

Curative Materials that react with an isocyanate prepolymer to produce the final elastomer.

Cycle time The amount of time required to complete a molding cycle, including mold preparation, insert loading (when applicable), release agent application, mixing and dispensing of components, reaction and demolding.

Cyclone A tapered piece of equipment to separate fine and course material by centrifugal spinning.

Degassing Also known as deaeration or vacuuming, removing air from a liquid material.

Demold The process of removing a specimen or cast from a mold.

Density The mass per unit volume of a material.

Dew point The temperature of a surface at a given ambient temperature and relative humidity at which condensation of moisture will occur.

Diamine Organic compound containing two NH_2 groups.

Dielectric strength The measure of polyurethane's ability as an insulating compound to resist the passage of a disruptive discharge produced by an electric potential.

Diisocyanate A chemical compound, usually organic, containing two reactive isocyanate groups. Isocyanate groups consist of a nitrogen atom bonded to a carbon atom bonded to an oxygen atom, $-N=C=O$. It is often abbreviated to isocyanate or iso in the polyurethane industry.

Dike Low enclosure around tanks or drums to prevent escape of liquid (bund).

Dumbbell A flat sample having a narrow straight central portion of uniform cross-section with elongated ends.

Durometer An instrument for measuring the hardness of rubber and plastics.

Elastomer A term often used for rubber and similar polymers.

Elongation (E@B %) Extension produced by a tensile stress.

Elongation ultimate The elongation at the time of break.

Endothermic A chemical reaction that absorbs heat from the environment around it.

Epoxy Synthetic resin compounds, usually thermosetting, that are capable of forming tight cross-linked polymer structures characterized by toughness, strong adhesion and high corrosion and chemical resistance. Used for making casts, coatings and high-strength adhesives.

Exotherm The liberation of heat during the course of a chemical reaction.

Filler A solid compounding material, usually fine, that may be added to a polymer for technical or economic reasons.

Filler, inert A filler having no reinforcing effect.

Flame retardant An added substance that inhibits the initiation and/or spread of flame and/or amount of smoke generated during combustion.

Flammable limits The concentration of flammable vapor in air, oxygen, or other oxidant that will propagate flame upon contact when provided with a source of ignition. The lower explosive limit (LEL) is the concentration below which a flame will not propagate; the upper explosive limit (UEL) is the concentration above which a flame will not propagate. A change in temperature or pressure may vary the flammable limits.

Flash The excess material protruding from the surface of a molded article at the mold junctions.

Flash point The lowest temperature at which a flammable liquid will give off enough vapor at or near its surface to form an ignitable mixture with air.

Flexural modulus The ratio, within the elastic limit, of the applied stress on a test specimen in flexure to the corresponding strain of the specimen.

Flexural strength The maximum stress required to be applied to the center of a test specimen (a bar-shaped test specimen that is positioned on two supports, one on either side of the load) required to crack or break the specimen.

Free isocyanate A measure of the free, unreacted diisocyanate monomer present in an isocyanate prepolymer. Low free isocyanate prepolymers are desirable as they contain very low levels of volatile isocyanate, making them inherently safer for the end user.

Gel time For polyurethanes, the interval of time between mixing together the polyol and diisocyanate or prepolymer and curative and the formation of a non-flowing, semi-solid, jelly-like system.

Glass transition temperature (Tg) A reversible change that occurs when plastic is heated to a certain temperature range, characterized by a rather sudden transition from a hard, glassy, or brittle condition to a flexible or elastomeric condition.

Green strength The rate of development of initial strength properties of polyurethane, soon after reaction completion. Rapid green strength development is desirable in molding operations, as it allows early removal of the molded part from the mold, thus decreasing molding cycle times and increasing the productivity of the molding operation.

Hard segment One of the two phases that makes up polyurethane. The hard segment is composed of polyisocyanates and chain extenders. The hard segment controls many of the polyurethane properties such as tensile, tear strength, hardness, compression set, abrasion/erosion resistance and thermal stability.

Hardness The resistance to indentation as measured under specified conditions.

Hazardous goods A classification for chemicals that describes the Workplace Health & Safety issues associated with its handling by end users.

HDI An abbreviation for hexamethylene diisocyanate.

Heat buildup The temperature rise within an elastomer due to hysteresis. In many end use applications, an elastomer can be subjected to repeated cycles of deformation-relaxation. As this occurs, friction between the elastomer molecules generates heat. As elastomers have relatively poor thermal conductivity, the heat generated builds up over time, progressively increasing the internal temperature of the elastomer. If the temperature increases over 70 °C, the elastomer's physical properties can begin to reduce. Design of the elastomer part can play an important role in minimizing the effects of heat buildup.

HMDI Abbreviation for hydrogenated diphenylmethane-4,4-diisocyanate.

Homogeneous A term describing the uniform composition of a material. It can be used to describe complete and thorough mixing of polyurethane or uniform physical properties throughout a cured polyurethane.

Hot cure Casting process for the production of high-performance elastomers in which the elastomer is mixed, dispensed and cured at elevated temperature.

Hydrolysis The effect experienced mainly by polyester-based polyurethanes, where prolonged contact with water or water-based liquids causes breakdown and failure of the polyurethane.

Hydrophilic Water-soluble or water-attracting molecules or systems that strongly interact with water.

Hydroxyl group The combined oxygen and hydrogen radical (-OH) that forms the reactive group in polyols.

Hydroxyl number The number of milligrams of potassium hydroxide (KOH) equivalent to the hydroxyl content of 1.0 g of polyol.

Hygroscopic A material that absorbs moisture readily.

Hysteresis The ability of polyurethane to absorb and dissipate energy due to successive deformation and relaxation. A measurement of the area between the deformation and relaxation stress-strain curves.

Inert gas A gas that exhibits great stability and extremely low reaction rates under normal temperature and pressure conditions, for example, nitrogen, argon and helium. Nitrogen is commonly used in polyurethane processing.

IPDI An abbreviation for isophorone diisocyanate.

Isocyanate The name of the chemical group containing a nitrogen, carbon and oxygen atom (-NCO).

Isocyanate decontaminant A liquid material consisting of 90% water, 8% concentrated ammonia and 2% detergent, used to neutralize isocyanate spills and decontaminate used isocyanate drums before disposal.

Knit line A visible line seen in a molded polyurethane part that indicates where two flows of reacting polyurethanes have come together and joined. Often caused by flow turbulence, incorrect injection point placement or poor polyurethane flow. Depending on the severity, the knit line can be a weak point in the molding that may cause future delamination, structural failure or poor dimensional stability.

Master cast A flawless cast, that is set aside to take the place of the specimen, should it need to be molded in the future.

Master pattern A flawless pattern similar in function to a master cast. The pattern preserves the detail of the mold as well as the detail of the specimen. There is one master pattern for each side of a mold. Therefore, a two-part mold would have two master patterns.

MDI An abbreviation for diphenylmethane-4,4'-diisocyanate.

MEK Methyl ethyl ketone.

MOCA (MBOCA) Trade name for methylene bischloroaniline, which is a widely used curing agent for hot cure polyurethane elastomers. There are ongoing debates about health risks associated with its use, as it is a suspected carcinogen.

Modulus The tensile stress at a given elongation.

Mold A manufactured cavity that preserves a negative impression of a specimen that can be filled with a foam or casting compound to produce a specimen replica. it can be manufactured from a variety of materials, depending on the production requirements, such as steel, aluminum, polyurethane, epoxy, FRP, silicone rubber or latex. It can be manufactured in "one piece" or in multiple interlocking pieces. Multi-piece molds are used when the cast has a complex shape or undercuts that would make demolding from a one-piece mold difficult or impossible.

Mold release agent A lubricant that prevents the casting from adhering to the mold.

Molding The process of forming a material to a desired shape by flow induced by force after the material is placed in the heated mold cavity.

Molecule A group of atoms held together by chemical forces. Each type of molecule is composed of a specific arrangement of atoms. It is the smallest quantity of matter that can exist by itself and retain the properties of the original substance.

Monomer The smallest repeating structure of a polymer. Usually a relatively simple compound containing carbon, which can react with itself or with other molecules or compounds to form a polymer.

MSDS (Material Safety Data Sheet) A document required by law to be supplied to a user of a hazardous product, to assist the user in safe use and handling of the hazardous product. The MSDS has now been replaced by a Safety data sheet (SDS).

MW Abbreviation for Molecular Weight. The sum of the atomic weights of all the atoms that constitute a molecule.

NCO Nitrogen Carbon Oxygen (the isocyanate group).

NDI The abbreviation for naphthalene diisocyanate.

Nonreactive Does not react in the situation.

On cost Noncapital or raw material cost associated with production.

Open pour The process of filling a mold by pouring polyurethane directly onto the lower surface of an open mold.

Optimum cure The state of cure at which a desired property value or combination of property values is obtained.

Orange peel (alligator skin) Term used to denote a typical surface texture caused by the settling of fillers from a batch during casting.

Over cure The state of cure beyond the state of optimum cure.

Part line The surface at which the various pieces of a mold come together when reassembled. The part lines of a mold determine the flash lines of the casts produced from that mold.

Pigment A powdered or liquid substance used in resins that imparts coloration to the cured item. Can be organic or inorganic.

Plasticizer A compounding material used to enhance the deformability of a polymeric compound.

PLC Programmable logic controller.

Polycaprolactone A polymer made by ring opening caprolactam.

Polyester polyol A chemical building block. The polyester provides good solvent resistance and good mechanical properties in the final polyurethane.

Polyether A chemical building block. The polyether provides good resilience, hydrolytic stability, mechanical properties and cost advantages.

Polyisocyanurate A polymer containing multiple isocyanate-to-isocyanate bonds.

Polymer A material consisting of molecules characterized by the repetition of one or more types of chemical units (poly = many, mer from monomer - mono = one).

Polyol An organic compound having more than one hydroxyl (-OH) group per molecule. The term includes monomeric and polymeric compounds containing hydroxyl groups such as polyethers, glycols, glycerol and polyesters.

Polyurea A polymer containing the urea group NH-CO-NH-.

Polyurethane A polymer containing the urethane group NH-CO-O-.

Porosity The presence of numerous small cavities.

Post cure Heat treatment to which a cured or partially cured thermosetting plastic or rubber composition is subjected to enhance the level of one or more properties.

Pot life The period of time during which a reacting thermoset plastic remains suitable for processing after mixing with a reaction-initiating agent.

PPDI An abbreviation for para-phenylene diisocyanate.

PPG Polypropylene glycol.

Prepolymer The product from reacting a polyol with an isocyanate.

Processability The relative ease with which a raw or compounded polymer can be handled.

Processing aid A compounding material that improves the processability of a polymeric compound.

PTMEG Polytetramethylene glycol.

PU or PUR An abbreviation for polyurethane.

Purging Process of expelling an unwanted gas or vapor from a system through the introduction of a different gas or vapor until all traces of unwanted gas or vapor have been removed.

Recovery The degree to which an elastomeric product returns to its normal dimensions after being distorted.

Registration The active alignment of the various pieces of a mold by tabs or locator pins, so that no portion of the mold will stray from the position it was in when it was initially formed around a specimen.

Relative humidity The ratio, expressed as a percentage, of the amount of moisture the air actually contains to the maximum amount it could contain at that temperature.

Release agent A substance put on a mold surface or added to a molding compound to facilitate the removal of the molded product from the mold.

Resilience The ratio of energy output to energy input in a rapid full recovery of a deformed specimen.

RIM Abbreviation for Reaction Injection Molding. The RIM process involves the rapid metering and mixing of polyurethane reaction ingredients, followed by their injection into a mold. It allows the rapid production of molded polyurethane components.

SDS A document required by law to be supplied to a user of a hazardous product, to assist the user in safe use and handling of the hazardous product. Replaced MSDSs.

Sealants A liquid, paste, coating or tape that fills holes, joints or gaps between mating surfaces, stopping leakage of gas, liquids or solids.

Set Strain remaining after the complete release of the force producing the deformation.

Shore hardness The hardness of a material as determined by either the size of an indention made by an indenting tool under a fixed load, or the load necessary to produce penetration of the indenter to a predetermined depth.

Shot The accurate dispensing of a precalculated quantity of mixed polyurethane from a polyurethane dispensing machine. The shot can be expressed in seconds or by weight of polyurethane dispensed.

Shrinkage A measure of a material's reduction in size after setting or curing. Usually expressed as a dimensionless ratio of the amount of shrinkage over a unit length. Dimensionally stable materials have shrinkages very close, or equal, to 0.

Silicone rubber A two-component synthetic rubber capable of curing at room temperature by chemical means into a solid elastomer. Commonly used as a mold-making compound where a soft, pliable mold is required. Silicone rubbers offer advantages of very low hardness, high flexibility and self-releasing properties but have the disadvantages of high cost, high viscosity and low strength, which limit their application and longevity.

Soft segments One of the two phases that make up polyurethane. The soft segment is composed of long-chain polyether or polyester polyols. The soft segment controls many of the polyurethane properties such as tensile and tear strength, hydrolysis and chemical resistance, glass transition temperature and flexibility.

Sprue The opening, or hole, through which the casting medium is poured into some molds. The term also refers to the waste material that hardens in the opening and often adheres to the cast. Alternately, the term can also apply to the piece of material that the mold maker originally places on the specimen to form the opening in the mold.

Strain The unit of change, due to force, in the size or shape of a body referenced to its original size or shape (e.g., elongation).

Stress The intensity, at a point in the body, of the internal forces that act on a given plane through the point. (This is expressed as a force per unit area for example, pounds per square inch or MPa).

Striations Lines of different color or intensity that are evident both in the unpigmented or pigmented polyurethane mixtures.

Swelling The increase in volume of a specimen immersed in a liquid or exposed to a vapor.

Taber wear index The ability of a material to withstand mechanical action such as rubbing, scraping or abrasion, that tends to progressively remove material from its surface. Usually expressed in milligrams loss per number of cycles per given load.

Tan delta(δ) The viscous modulus/elastic modulus.

TCP Tricresyl phosphate.

TDI An abbreviation for toluene diisocyanate.

Tear strength The maximum force required to tear a specified specimen, the force acting mainly parallel to the major axis of the test specimen.

Tensile modulus The ratio of stress to corresponding strain below the proportional limit of the material. Normally expressed in MPa (megapascals).

Tensile strength The maximum tensile stress applied during stretching a sample to break.

Thermoplastic A resin or plastic compound that can be repeatedly softened by heating and hardened by cooling. Examples of thermoplastics are acetal, acrylic, chlorinated polyether, fluorocarbons, polyamides (nylons), polycarbonate, polyethylene, polypropylene, polystyrene, some types of polyurethanes, and vinyl resins.

Thermoset A resin or plastic compound that in its final state is substantially infusible and insoluble. They cannot be repeatedly softened by heating and hardened by cooling. Examples of thermosets are epoxy, phenol-formaldehyde, some types of polyester, some types of polyurethane and urea-formaldehyde resins.

Thixotropy A flow characteristic of certain fluids where a decrease in viscosity of the fluid occurs when it is stirred at a constant or increasing rate of shear. When the stirring or shearing is discontinued, the apparent viscosity of the fluid gradually increases back to the original value. Changes in both directions are dependent on time as well as shear.

Toxic A substance that has the ability to produce injurious or lethal effects through its chemical interaction with the body.

TMXDI An abbreviation for meta-tetramethylxylylene diisocyanate.

TPU Thermoplastic polyurethane.

Triol A polyol containing three reactive hydroxyl groups.

Trowellable Capable of being applied using a trowel and having thixotropic properties.

Turbulence Used to describe an erratic, tumbling flow of polyurethane foam or elastomer through a mold or cavity during filling. Usually caused by poor mold design, incorrect location of the polyurethane injection point, obstructions in the mold or cavity or poor polyurethane flow properties of the polyurethane formulation.

Ullage Space left in the top of a drum to allow for expansion of contents.

Under cure The state of cure between the onset of curing and the state of optimum cure.

Undercut A part of a mold, cast, or specimen that deviates from a sloping or vertical surface and turns back onto itself.

Urethane The chemical group NH-C=O-O-.

UV stabilizers Additives incorporated into a polymer that prevent or slow the degradation of the polymers caused by exposure to light.

Venting The displacement of air from the mold cavity as the cavity is filled by polyurethane. Venting normally occurs through small holes or seams in the mold located at strategic positions around the mold to ensure that all air is vented. When the mold is completely air free and polyurethane filled, a small amount of polyurethane also vents, further ensuring a completely air-free filling of the mold.

Viscosity The resistance of a material to flow under stress.

Water absorption The amount of water absorbed by a material under specified conditions.

Index